AF760179

How the Math Gets Done

How the Math Gets Done

Why Parents Don't Need to Worry about New vs. Old Math

Catheryne Draper

ROWMAN & LITTLEFIELD
Lanham • Boulder • New York • London

Published by Rowman & Littlefield
A wholly owned subsidiary of The Rowman & Littlefield Publishing Group, Inc.
4501 Forbes Boulevard, Suite 200, Lanham, Maryland 20706
www.rowman.com

Unit A, Whitacre Mews, 26-34 Stannary Street, London SE11 4AB

Copyright © 2017 by Catheryne Draper

All rights reserved. No part of this book may be reproduced in any form or by any electronic or mechanical means, including information storage and retrieval systems, without written permission from the publisher, except by a reviewer who may quote passages in a review.

British Library Cataloguing in Publication Information Available

Library of Congress Cataloging-in-Publication Data Available

Includes bibliographic references.
ISBN 978-1-4758-3422-2 (cloth : alk. paper)
ISBN 978-1-4758-3423-9 (pbk. : alk. paper)
ISBN 978-1-4758-3424-6 (electronic)

™ The paper used in this publication meets the minimum requirements of American National Standard for Information Sciences—Permanence of Paper for Printed Library Materials, ANSI/NISO Z39.48-1992.

Printed in the United States of America

Dedicated to all the students who taught me how to teach them.

Eric	Connie	Mary	Jena	Elizabeth	Mandy	Evan
Rachel	Katie	Jared	Maggie	Nathan	Sean	Lindsey
George	CARLOS	Roger	Kaylee	Heather	Jo	Reiss
Jacob	Ethan	Hannah	Lexi	Mora	Deborah	Libby
Brendan	Robbie	Jonathan	Olivier	Christine	Ed	Kim
Jack	Emma	Neil	Madeline	Bobby	Kendra	Nicole
Joey	Shane	Walker	Grace	Francis	Brian	Dee
Kristen	Jaclyn	Tom	Kajsa	Nicole	Seamus	Colleen
Adam	Joshua	WILL	Bianca	Jay	Chris	Lisa
Andrew	Sydney	Amber	Melissa	Janis	David	Traci
RICHARD	Jane	Becky	Jason	Jacqueline	Peter	WILL
MADELINE	Nell	STACEY	JENNIFER	Jason	Ralph	Holly
Talia	Ren	Mitch	Wendy	Rachel	Carlo	Leigh Ann
Ted	Joan	Ben	Richard	Teddy	David	John
Danielle	Kathleen	Rob	EJ	Randy	Raymond	Beth Ann
Kelly	[illegible]	[illegible]	[illegible]	[illegible]	Phillip	Stephani
Cody	[illegible]	[illegible]	[illegible]	[illegible]	Chris	Kelly
Connor	Don	JESSICA	Cathy	Mark	Jeffrey	DEBORAH
Nate	Liv	Erik	Roberta	Catie	Dan	Chloe
Andy	Whitney	Teresa	JAMES	Doug	Shawn	Mallory
Caitlin	Max	Maggie	Kevin	Derek	Brandon	Terri
Russell	Matthew	CHARLES	Margaret	Lauren	Tanun	William
Rick	Geoffrey	KATY	Deirdre	Joe	Sean	Devan
Jessica	Chandra	ANNA	Megan	Lisa	Steven	Jan
Bradley	Timothy	Paul	Nancy	Dana	C. J.	Lee
Doug	Elizabeth	Brendon	Ricky	Diane	Jean	Shane
Ken	Connor	Marie	Peter	Roseanne	MATHEW	Robert
Katie	Angela	John	Rachel	HEIDI	Taylor	CAMILLE
Ryan	Betty	Joanni	Mallory	LORY	Jeff	Molly

Contents

Foreword

> I never got a pass mark in math. . . . Just imagine—mathematicians now use my prints to illustrate their books.[1]
>
> —M. C. Escher

I barely passed several of my own math classes. I can take comfort that I did better than my husband, who *tied* for the lowest grade in a calculus class of several hundred students—with a score of *sixteen.* It is hard to believe I am writing a foreword to a book about math! Let's face it, when the topic of math comes up, for many of us our response is "I am not good at math" or "I hate math." Just google the subject, "I am not good at math" and more than 6,910,000 responses come up. Google "I hate math" and you will get another 2,520,000 responses.

It is clear that a large number of us somehow end up believing that we are not good at math, and, as a result, do not like math. Even M. C. Escher's teachers did not see a future genius. It is his work that made me stop and think that most of us are actually quite good at math, if not great at it. We use it when we design and make quilts, measure for recipes, build houses, and design and build furniture, just to name a few examples.

Just think for a moment the role that math plays in our lives. There are really not many things that we do each day that do not in some way have something to do with math. I recently read an article that pointed out, as other articles have, that without math, there would be no science. If you think about it, there would be no literature either. The early *Mesopotamian cuneiform* writing system was associated with an earlier system of numerals. So clearly, we need to get past thinking that we are not good at math.

I have two children, each of whom struggled with math. A friend referred me to Cathy Draper; and thank goodness for that referral. As you read *How the Math Gets Done*, you will learn how to help your children to never feel or say they are not good at math. You will also, no doubt, learn that you were probably very good, if not great, at math! I have had the privilege of watching my children being tutored by Cathy, and to see them becoming engaged to the point where I might even dare to say that they are learning to love math.

All (or at least nearly all) of Cathy's techniques and secrets are in the chapters of this book and will help you to help your child navigate the road of mathematics. It is like having a private lesson with Cathy without having to leave your home. As you read this book, you will find yourself saying, "If only I had been taught this way in school." I know I heard myself say that many a time while in the room during my children's time with Cathy.

I am excited that you too will have at your fingertips the knowledge and means (I hope you get the math pun) that my children learned directly from Cathy. *How the Math Gets Done*, along with the other books in this series, *Winning the Math Homework Challenge* and *User-Friendly Math for Parents* will have you throwing away your preconceived notions that math is dreary and difficult if not impossible.

Cathy's secret is making you understand that math is the most creative subject! Yes, through this book, you and will learn to put aside what turned you away from math, and begin to see math as a search for patterns and a way to solve puzzles through deductive thought. Better yet, you will be able to pass on this new way of thinking, learning, and approaching math to your child. Remember, two plus two may equal four, but a love of math can be infinite! *How the Math Gets Done: Why Parents Don't Need to Worry about New vs. Old Math* will help you and your children get there.

Robyn Frost

NOTE

1. Escher, M. C. "M. C. Escher Quotes." AZ Quotes. www.azquotes.com/author/4555-M_C_Escher.

Preface

How the Math Gets Done: Why Parents Don't Need to Worry about New vs. Old Math explains how the presumably "new math" is not new at all. The arithmetic procedures for adding or long division may be different, but the mathematics that makes these procedures actually work is the same. Like *Winning the Math Homework Challenge* and *User-Friendly Math for Parents*, this book continues to tell real stories about how children think as they learn about calculating and making new numbers by adding them and multiplying them.

Seeing the mathematics of calculating through your child's eyes, and understanding *why* they may see it that way, is the purpose of *How the Math Gets Done*. This book follows the imaginative and creative learning premise for learning math through visual strategies featured in *Winning the Math Homework Challenge* and *User-Friendly Math for Parents*. As you read these children's stories, be prepared to get excited about understanding your own feelings and similar experiences. Also, prepare to be challenged to understand children's descriptions and explanations as seen *through their eyes*.

We don't know what we don't know. In this respect, adults are no different than young learners. Parents want to help their child understand. Parents who are Math Aficionados want their children to understand what is so very clear to them and have little clue about what is so confusing. The Math Avoider parents want their children to know more than they themselves were able to learn. Both types of parents have an equal passion to help. Their children are caught in between.

My hope is that these stories can help you, the parent, see the arithmetic of mathematics through your children's eyes, both the clarity of interpretations and the confusion in the slash marks of "borrowing." Armed with this new insight about their confusions, you will be able to talk differently with your children about why adding works the way it does and why the organization for all of the operations is important. If you are a Math Avoider, you might recognize yourself in these stories. If you are a Math Aficionado, you might get a glimpse of the reasons behind your children's dilemmas.

Visual-spatial children learn differently than auditory-sequential children. Visual-spatial children work best when they know "why," and they do not react well to the details of step-by-step explanations without knowing the larger scope of the "big idea." The explanations in *How the Math Gets Done* capitalize on the spatial strengths of these visual-spatial and visual-mechanical learners. Images and diagrams, overall "big idea" concepts, creative explanations by other children, organization, and analogies and metaphors are used as examples in the explanations instead of simply providing definitions and rules.

The children featured in *How the Math Gets Done* started their turnaround from failure to success when the visual materials started making addition, subtraction, multiplication, and, yes, long division make sense to them. They moved the materials around, arranged, rearranged, and examined them from different perspectives. They asked questions about the relationships *in the materials*. As students tried out new arrangements, a subtle yet effective psychological shift occurred. Their aversion to the math transferred *from their core identity to the materials*. The math relationships and any associated emotions were now in the materials, where they should have been all along. With this transference came the ability to work with math relationships externally instead of as an internal inadequacy.

These children also started to take responsibility for their learning, slowly and tentatively at first, by risking some opinions and trying out some imaginative interpretations. Long division ceased to be enigmatic; multiplication with mixed numbers was just a rearrangement of blocks. The children were beginning to feel like they may indeed *have some control* over their math learning! As their understanding improved, so did their math voice. Isn't that the real purpose behind learning?

Introduction

How the Math Gets Done: Why Parents Don't Need to Worry about New vs. Old Math provides a roadmap to understanding what the symbols for math operations (add, subtract, multiply, and divide) *really* mean, what the clues are to interpret these symbols, and a kind of short story of how they evolved over time. Math doesn't change and it is not "new." It is the same as it has been for millennia, only expanded with new innovations like spherical geometries or chaos theory and fractals, *but not new*. The mathematics diagram ("quadrivium" for the history buffs) in Plato's *Republic* includes the same arithmetic that we have today.[1]

Newer and different methods or explanations may "come on the scene" every decade or so and get politicized as yet another "new math," but none of these "new math" initiatives represent the mathematics; they represent newer opinions of better ways to teach or to compute with newer versions of the old procedures. Each new initiative brings *new insight and ideas* to better help children *understand the same math*. Math aficionados are allowed to have their opinions, just as the children in the stories in this book are allowed to express their interpretations. "Better" explanations are opinions, and may not be "better" for everyone.

This book is not a math-made-easy book. Math isn't easy, but it *is not* as strange and mystical as its reputation suggests to Math Avoiders. Math isn't flat after all, as one student observed; math has texture that you can grab onto. Thinking mathematically is the best mathematics, and learning the best mathematics gives the satisfaction of winning a marathon of long-term connections and patterns, not running a sprint of shortcuts and sound bites.

"It was good enough for me" isn't working as a justification to keep the instruction the same. Same instruction while expecting different results has a dubious reputation. Researchers studying visual learners have found that the majority of the learning population has visual-spatial preferences for learning, a quantity that certainly requires some consideration for instructional changes. Drilling and memorization still exist, sometimes out of pure frustration, and neither is very helpful to learners, especially visual learners. It is time to listen, listen beyond just the hearing, to understand what children are "understanding" about math.

The craft of mathematics can get lost when children are just "doing the practice examples" on endless amounts of worksheets. Children deserve to have the opportunity of experiencing the benefits of exploring new ideas as earlier mathematicians have done, like Bombelli with his "wild thoughts" about imaginary numbers or Chuquet's discovery of those "absurd" negative integers. Making mistakes and learning from mistakes are critical to productive learning. Children will benefit from the new information that results from a "no" response, and they will learn that "no" is central to k*no*w, both literally and intellectually.

All children should have the opportunity to find their math voice, including the visual-spatial learners and any of the other frustrated and bewildered children who are currently learning math. Frustrated parents also deserve to discover and use their math voice. Math is about thinking and communicating through a math language. That is why *How the Math Gets Done* tells children's stories about their understandings and misunderstandings, their successes and their mistakes, and how they learned to decipher the enigmatic squiggles of those verbs called operations.

How the Math Gets Done compares the old and the new methods for math procedures from a "big idea" perspective by organizing the information in four sections: "Definition," "Organization," "Relationships and Patterns," and "Connections." Each section contains three chapters that clarify the issues related to each "big idea" section. The conclusion offers parents even more hints and guidelines to help their children through this "math country" of procedures for calculating in math.

The first chapter in the "Definition" section reflects on how children interpret what we are "teaching" them about the small symbols for math operations that simply make new kinds of numbers. Chapter 2 talks about the verbs of the operations, which symbols have "put-together" verb interpretations and which ones have associations with "take-apart" verbs. The focus is on the *actions* involved with these four math operations. Chapter 3 brings rote memorization into the spotlight and speaks to how well it actually works for the long term.

The three chapters in the "Organization" section explain why no single procedure is best for everyone. Chapter 4 describes how important classification and organization are to the operations of adding, subtracting, multiplying, and dividing. Chapter 5 discusses children's imaginative interpretations of the operations. Chapter 6 looks at the "old way" format and compares that traditional format with several "new ways" to add, subtract, multiply, and divide. These comparisons show how the different procedures simply place digits in different locations, but *do not change the place value categories*.

The "Relationships and Patterns" section includes three chapters describing the importance of patterns and how the patterns tie the relationships together in the craft and design of real mathematics. Chapter 7 describes a diagonal and vertical viewfinder that can help children focus on specific math number patterns in the multiplication and addition tables. Chapter 8 expands on the diagonal and vertical viewfinders with two more, the quarter-block viewfinder and the triangle-block viewfinder. Chapter 9 is about functions, the predictable equation relationship that reflect the patterns that are so important to mathematics.

The "Connections" section outlines three examples for connecting math topics that can jumpstart new searches. Chapter 10 describes how an ancient area design applies to all those numbers in math class. Chapter 11 is about ratios, proportions, and rate of change and their connections to each other and to previous topics. Chapter 12 brings the arithmetic and patterns from earlier years into upper level generalized algebraic expressions.

The math topics in the chapters are associated with grade levels as well as with how the same topic appears and reappears in later mathematics or within other contexts. The grade levels are guidelines only, since curricula differ among schools, states, and regions. The idea is to explain the topic first using visual-spatial strategies and then show parents how the topic recycles throughout mathematics in different forms. Learning the math concept behind the topic appropriately *the first time* means that you do not have to relearn it later.

The bibliography is not limited to the sources quoted or referenced in this book; the list embraces a broader range of sources that can be helpful for parents to investigate further in order to learn more about the ideas and activities in this book. Neither is it an exhaustive list, as that would possibly be as large and as extensive as an unabridged dictionary. Many of the early visionary math educators' references are included in this book because many of the current visual, imaginative, and creative activities can be traced back to their work.

NOTE

1. Turnbull, *The Great Mathematicians*, 11.

Part I

DEFINITION

> I had been to school most all the time and could spell and read and write just a little, and could say the multiplication table up to six times seven is thirty-five, and I don't think I could ever get any further than that if I was to live forever.[1]
>
> —Huck Finn

Arithmetic is about calculations; mathematics is about the system *that makes the arithmetic work.* When the two are confused, complications abound under the umbrella of all of those errors or so-called "careless mistakes," one-time-applications called "sound bites," "short cuts," misdirections, and those three-minute multiplication drills, recitations, and tests (poor Huck didn't get any farther than six times seven—and the accuracy of his recitations are clearly dubious). In desperation, rote memorization takes hold with an unrelenting survivalist grip. Just "saying the tables" and the other umbrella survival strategies interrupts the understanding of the calculations.

Did you ever understand *why* mantras like "invert and multiply" worked? "The sad thing is not that this doesn't work, but that it does . . . in the short term."[2] Did *you* ever ask why something worked, only to be told to "just do it," so you stopped asking why or even thinking about asking? Is your child in the same predicament, getting totally exasperated when you do try to explain? Are you guilty of saying "just do it" to your own child because you yourself never got a reasonable answer and still *really don't know why*?

This "Definition" section answers many of those "why" questions with plain English descriptions, pictures, diagrams, organization priorities, patterns, relationship translations, and several analogies and metaphors, all designed to provide an alternative perspective to survivalist tactics like rote memorization. When there is only "rote memorization" without any meaning, then any efforts toward building future understanding on previously learned information will be useless. Besides, our visual-spatial children, who make up 63 percent of the total learning population, do not learn or remember through practice and memorization.

Chapter 1 reflects on how children interpret what we are "teaching" them about the small symbols for math operations that simply make new kinds of numbers. Chapter

2 talks about the verbs of the operations, which symbols have "putting together" verb interpretations and which ones have associations with "taking apart" verbs. The focus is on the *actions* involved with these four math operations. Chapter 3 brings rote memorization into the spotlight and speaks to how well it actually works for the long term. Isn't long-term learning the actual goal?

The stories in this section highlight children's deeper thinking observations, like eight-year-old Guy, who quietly pondered over some cheese pieces before announcing, "I have a theory. One-third plus one-half of one-third is one-half." Guy's mom calculated the fractions with her school-learned rituals before she agreed, then she asked Guy to test his idea with numbers other than three. She asked, "Will this work for other fractions?" Observations are about thinking mathematically; testing them out is real mathematics!

You will read about a fourth-grade class that explored numbers by cutting out the number shapes as rectangles. In the process, they learned that most of the numbers could be cut out in different ways to make different-looking rectangles. They named these different rectangle arrangements "spatial number personalities." These fourth graders used their imaginations and ingenuity and made the math make sense to them!

NOTES

1. Twain, *The Adventures of Huckleberry Finn*, 33.
2. Sutton, "Remedies for the Pain of Math Anxiety," 7.

Chapter One

Math According to Jen, Bobby, and Others

> Whatever its essence, we recognize [mathematics] to be as free as the mind, as prehensile as the imagination.[1]
>
> —Edward Kasner and James Newman

Jen and Bobby invented their own "imagination grasping" explanations for the math symbols of calculations when the teacher's explanation, or their parents' interpretations, didn't make sense to them. Their *own* explanations, however accurate or not mathematically, were imagination grabbing *to them* (Kasner and Newman's prehensile). Listening, really listening beyond just hearing, to your child talk about what *they think* that the symbols mean provides insight into how easy it is to misunderstand such a small mark on the paper. This chapter shares children's stories so that parents can gain insight into how these symbols can be confusing.

These very small calculation symbols (+, –, ×, and ÷) represent very large mathematical ideas. The challenge for your child is learning about these very small yet powerful symbols at the same time that they are learning about those other small symbols called letters, spelling, and how to decipher both. The comments in figure 1.1 describe what children have said about these operation symbols, which to them can seem rather whimsical. Maybe your child has made these same or similar comments about the symbols. With your guidance and patience, your child can take the time to think and correct their own thinking with their one hundred billion neural connections.

JEN'S DEDUCTIONS

Jen, a second grader, interpreted the multiplication symbol "×" with her own explanation, "Turn the '+' and use the same numbers. If $2 + 2 = 4$ and $2 \times 2 = 4$ then $3 + 4 = 7$ means that $3 \times 4 = 7$." Jen had invented her own explanation *and* deduction for the symbol, only to learn that her pattern was not valid for numbers greater than two. Jen had recognized a pattern with the twos and had generalized it in a way that seemed very logical to her. She had used her visual-spatial pattern reasoning by focusing on

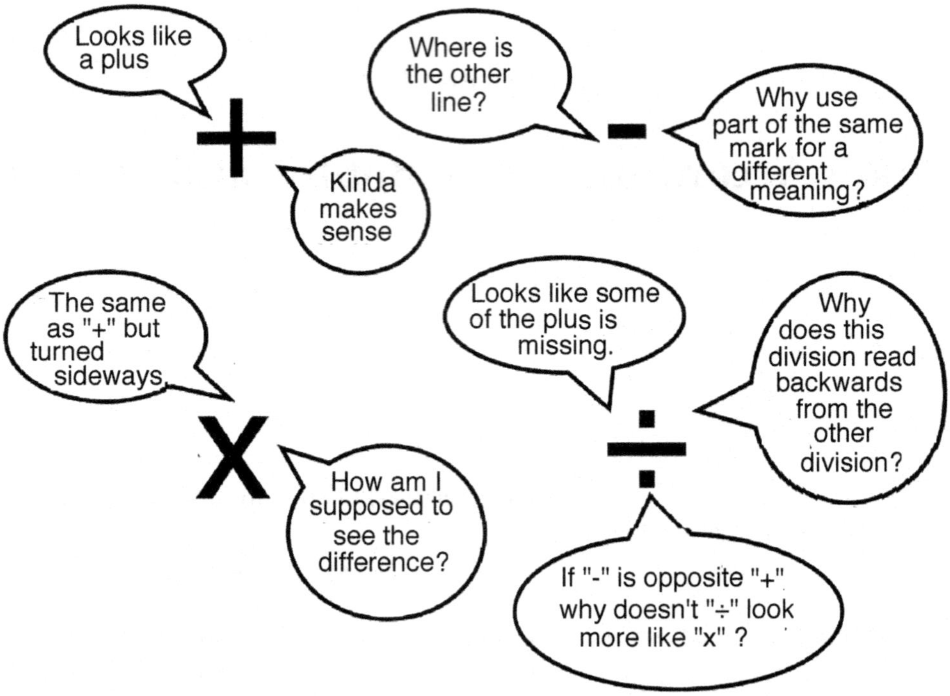

Figure 1.1. Children's comments.

what the symbols looked like. The *real trouble* for Jen (and others) is that this reasoning *was* valid based on the twos, giving her a false sense of clarity.

This search for mathematical clarity based on a single example can be addictive, and it lasts into the upper grades or even adulthood with more and more encounters with mathematical machinations of "imitative math" (the kind of math that replicates moving symbols but doesn't make sense). The jump to a conclusion based on a single example can give an adrenalin rush when "at last something *does* make sense" even if it is temporary. Appropriate mathematical generalizations *are* very satisfying (just ask Andrew Wiles, who completed the proof of a famous math theorem unsolved for three hundred years). A dependable proof depends on far more than one example.

The upper level versions of Jen's deductive generalizations that are based only on twos make their appearance with "if $2^2 = 4$, then $3^2 = 6$ and $4^2 = 8$" and so on. Gone unchecked, the deduction Jen made in the second grade will continue to reoccur in the upper grade levels with declarations such as "$^3/_2$ is the same as 3.2 because fractions are decimals." Statements like these depend on manipulating symbols without meaning and without number sense. Many times they are declared quickly to reduce the pain of not understanding. Thinking, understanding, wait-time for responses, and making sense of the symbols takes time.

Adult explanations trying to correct her thinking had not made any sense to Jen, so she had resorted to memorizing "math facts," a strategy that does not work well for visual learners. Jen had created her own reasoning based on how she "saw" the shape orientation of the symbols, and now her memorization efforts generated even more errors. Jen needed actual tangible models to test the fives, sevens, and other numbers to help her tell her side of the story *and reinvent her own deduction.* Her new descriptions were not like the ones in her book, but they did consistently generate accurate results. Math was now reasonable to her again.

Jen isn't the only one who thinks this way and who gets confused when a procedure works for twos but not other numbers. Unless her deduction is reconfigured *in her own mind,* Jen will continue generalizing procedures based on using twos, and the errors will happen again as "careless mistakes" in a wide variety of upper grade settings. Just as Guy's mom didn't "correct" Guy's theory about a fraction relationship based on threes, but instead asked for more theory tests with other numbers, you can help your child *reinvent their* deductions. When your child can self-correct their own thinking, the result is longer-lasting and more reliable learning.

BOBBY'S EXPERIENCE WITH MULTIPLICATION AND NUMBER ARRANGEMENTS

Bobby, a fourth grader, used linking cubes to show how numbers can be assembled into several different multiplication arrangements (also known as factors). His favorite arrangement was the long stick "Lightsaber" of *Star Wars* fame. Two examples of Bobby's arrangements, one for twenty-four and the other for thirty-six, are shown in figure 1.2. Both started with a long lightsaber and included all of the different ways that he could arrange the numbers. When Bobby wrote about these arrangements, he first described them as multiplication and then he confirmed his multiplication with addition. For Bobby, addition was more dependable.

Later, when Bobby was asked to use the cubes to show "six times five," Bobby promptly, and not surprisingly, made the arrangement shown on the left in figure 1.3. Seen through his eyes, this arrangement makes sense because that is the way it is written on a sheet of paper. Bobby is like many other children who respond to this request with this kind of arrangement. What *was* a surprise was the speed with which he changed his own mind! When asked the next question, "How many cubes will six times five be?" he immediately said "thirty" and then glanced at his own cubes. After an ever-so-dependable time-to-think pause, he mused, "It can't be."

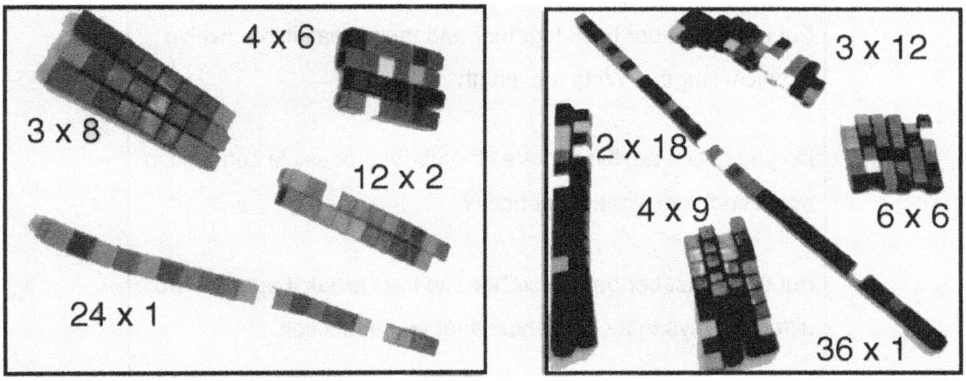

Figure 1.2. Bobby's arrangements for 24 and 36.

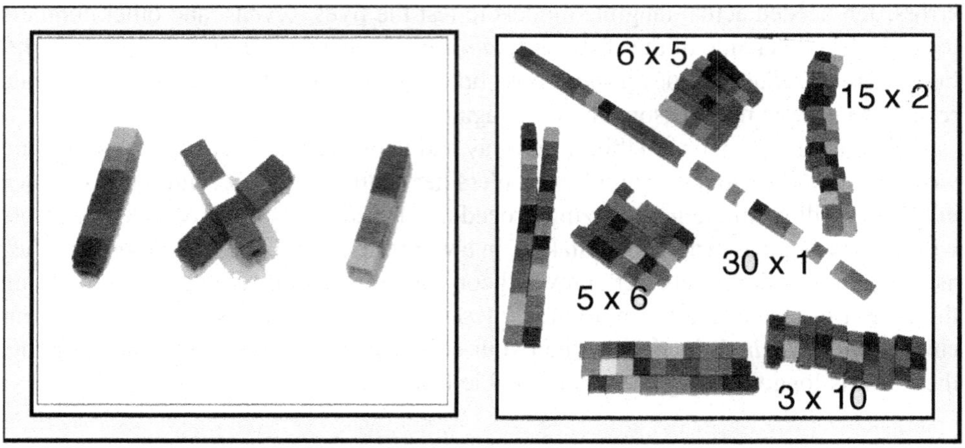

Figure 1.3. Bobby's "six times five" and corrections to make arrangements for thirty.

Bobby unlinked the cubes, collected more cubes for the necessary thirty cubes that he needed, and relinked his thirty-cube lightsaber. He was so delighted to have none left over after breaking his lightsaber into six sticks of five-cube lengths. By now he was on a roll because he had remembered his earlier arrangements. He proudly showed the five rows of six cubes and a rearranged six rows of five cubes. Then he made the rest of the arrangements for two times fifteen, three times ten, and what he called "their partners." Bobby, like Jen and Guy, made *his own adjustments to his own thinking* and not because he was "corrected."

Try This:

Use linking cubes to make the following arrangements. Count 48 cubes and link them into one long "lightsaber."

Break apart the long lightsaber into four identical shorter lengths. Write the length.

Put the lightsaber back together and then break it into twelve identical lengths. Write the length.

Do you notice anything interesting about a possible connection between these two experiences?

Put the lightsaber back together and then break it apart in two different ways that could show another connection.

Figure 1.4. Try This.

CHILDREN'S DESCRIPTIONS ABOUT MAKING SENSE

In a New Zealand study,[2] second graders were asked about the meaning of "times." One child described times as a counting idea. She said, "You count them out. Say, if it says two times three, you count the three twice . . . because the two says you do the three twice." Another child was asked if "times" meant the same as multiplication. The child responded that they were different because, "in multiplication, if you have a high number and want a low number . . . you get a multibyration [sic] equation." This child was possibly thinking about two separate but related ideas—multiplication and division—so putting them together made sense to her.

Still another child said, "Eight is larger [in $2 \times 4 = 8$] 'cause it is a higher number." Did this child mean that multiplication makes "higher" numbers? The eight is a higher number, but if this child believes that multiplication always makes numbers greater, then when fractions come into play, the child will become confused. Multiplying two fractions will generate a lower number amount. Oh dear, another disenchantment with math with another sound bite rule broken! Might as well have the conversation now about multiplication being about groups.

The interviewers in this study learned what many teachers and parents have already experienced—if the child has "insufficient concrete experiences, or concrete experiences devoid of understanding, [they bring] their own conceptual frameworks to bear." In other words, children tend to make up a very good story that makes sense to them, but may not necessarily make mathematical sense in the long run. It is very dangerous to require children to make sense out of nonsense—because they *will* make up a story, reasonable or not, and suffer any consequences later.

If you (or your child) are a member of the pushed-too-fast-too-soon club, then you probably have early memories about math never making much sense or you felt betrayed because it had what felt like secret rules. If you caught on eventually, or at least temporarily accepted the explanations as reasonable, then math may have made sense to you for a while, until the tsunami wave of new symbols that are introduced in upper-level grades started coming at you. If you were one of the ones who caught on early and even started playing with the symbols on your own, you probably never understood why your friends were having difficulty—to you it all made perfect sense.

Now imagine how your child feels (or how you felt) to have to write one hundred answers to one hundred problems and all one hundred problems have all sorts of strange symbols and you have to write all one hundred answers in just three minutes. If you haven't yet figured out the meanings of the symbols, then all of the symbols seem to blur together. One parent described this as feeling "like flying at Mach 3 speed into a black hole."

TREVOR AND JIM INVENT THEIR OWN MULTIPLICATION METHODS

A young third grader named Trevor had his own way of calculating and making sense of multiplication even before he was supposed to know how. His teacher had decided

Trevor's strategy:

\$2.80	\$2.80	\$2.80	\$2.80
x 31	x 10	x 10	x 10
	\$28.00	\$28.00	\$28.00

\$28 + \$28 + \$28 = \$84 then add another \$2.80

Figure 1.5. Trevor's work.

to not teach a lesson about a particular procedure; instead, she wanted her class to experience how good problem solving felt when they could solve the problems in their own way.[3] Trevor's solution in figure 1.5 incorporated what Math Aficionados call the distributive property, even before he knew that it was a property.

In fact, several of the students in this class had devised addition solutions for multiplication situations. Another problem was 75 × 15 and Jim, Trevor's classmate, wrote the seventy-five in a long list fifteen times. His list of the seventy-fives was a very long list, so young Jim wondered if he could make it easier for himself. He decided that he needed to figure out something that was shorter and neater than just adding up all those seventy-fives in the problem.

Jim's strategy:

75 + 75 + 75 + 75 + 75 + 75 + 75 + 75 + 75 + 75 + 75 + 75+ 75 + 75 + 75 = 1,125

Jim added the fives to get seventy-five. He wrote the five in the correct place value location for the seventy-five total for the fives and then he regrouped the seven from that seventy-five because he was comfortable with the place value organization. His next stages were as follows:

7 × 7 = 49 and 49 + 49 = 98 so 98 + 14 = 112. Jim's total is 1,125.

Can you figure out what Jim was thinking in his strategy? Don't forget about that regrouped counting by fives place value. Jim didn't.

This teacher knew the value of confirming accuracy without telling either Trevor or Jim or any of the others in her class what they were supposed to do to solve the problem. She also believed in ingenuity, like Guy's mom and Jen's teacher. Both Trevor and Jim had their own methods for solving their respective problems and they both had number sense, as well as a good dose of flexibility. These stories are reminders that mathematics should make sense *first*, and then conform to a more formal way of writing only if necessary.

KEEP IN MIND

Hearing what your child is really saying about the operations will give you more patience with your child's efforts. *Really* listening to your child can help you hear how *they* interpret operations so that you can steer them toward mathematical thinking. These kinds of math conversations will stop the futile explanation repetitions or the "starting back at the beginning" re-explanations of what they didn't understand the first, second, or even third time! Jen, Bobby, Trevor, and Jim, along with your child, have their own ways of thinking. It is more efficient *and effective* to *involve their thinking* as they learn, understand, and even invent their own strategies.

NOTES

1. Kasner and Newman, *Mathematics and the Imagination*, 359.
2. Katterns and Carr, "Talking with Young Children about Multiplication," 21.
3. Caliandro, "Children's Inventions for Multi-digit Multiplication and Division," 422.

Chapter Two

Math Grammar of Nouns, Verbs, and Stories

> "The symbols of combinations . . . [such as] addition . . . play a role similar to that of verbs."[1]
>
> —Philip J. Davis and Reuben Hersh

The symbols (+, –, ×, ÷) may be small, but they have very big meanings. The math operations for addition, subtraction, multiplication, and division are about actions with numbers. The math operations do the "work" and act on numbers. Verbs are about actions with nouns. While your child is learning about nouns and verbs, it makes sense to incorporate the same ideas into the math language using numbers as nouns and operations as verbs. The grammar of verbs and nouns as math operations and numbers is what word problems are all about. How all these ideas work together is what this chapter is all about.

Recognizing how verbs are used for the operations will help you explain these symbols to your child. In this chapter, you will learn what these small symbols *really* mean and how these real meanings as verbs can impact solving word problems as stories. The "putting together" verbs apply to addition and multiplication. The "taking apart" verbs are used with subtraction and division and are the inverse (opposite) operations of the "putting together" verbs. In fact, the inverse concept is a "big idea" in mathematics. Being flexible with these verbs with numbers puts success in mathematical thinking within easy grasp.

"PUTTING TOGETHER" VERBS FOR ADDITION AND MULTIPLICATION

Web diagrams are one of many kinds of graphic organizers that help to visually organize and relate concept clusters, such as story themes and history events. The web diagrams for the addition and multiplication "putting together" verbs are outlined in figure 2.1, showing several verbs and also including a caution about how addition and multiplication can have a different impact when using different kinds of numbers. The

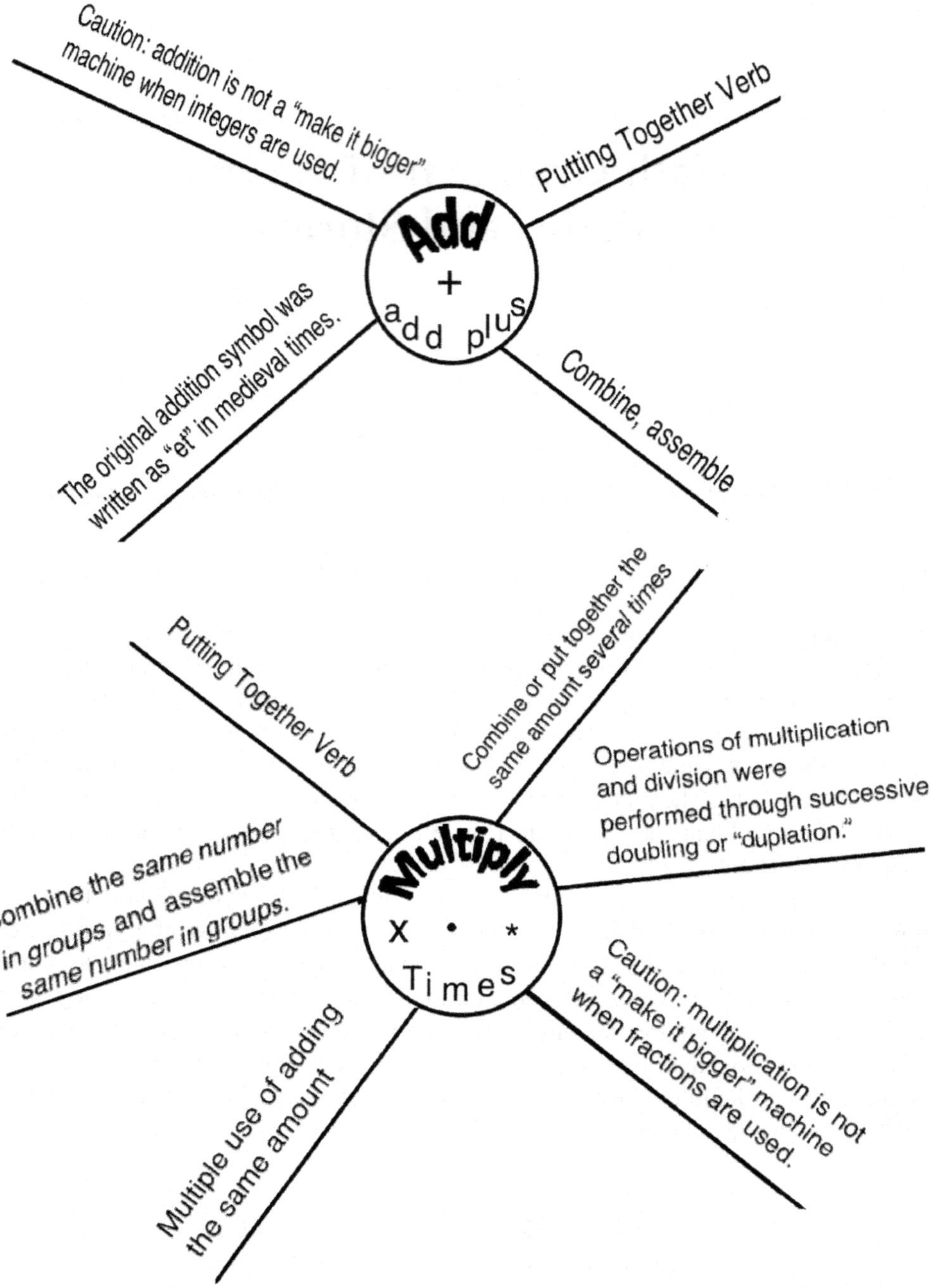

Figure 2.1. "Putting together" verb webs.

space between the spokes provides extra space for more notes to be included. These operations have a long history, dating as far back as "duplation,"[2] the Egyptian method for multiplication, and the medieval "et" for addition.

A second grader named Wendy was learning about multiplication. She did not understand the explanations, so she used her own intuition to explain what was going on with the numbers in a multiplication problem. To Wendy, the "three" and "four" are counted differently in "3 × 4," just as the "three" and "four" in "thirty-four" were counted differently. Wendy had intuitively connected two concept ideas: the idea of how digits are used in place value columns to the idea of how numbers are used differently in the multiplication that she was learning. In other words, Wendy's thinking was quite sophisticated.

Wendy's reasoning:

> In addition, both numbers always had some items to count to show the addition, but for a multiplication problem, one of the numbers in the problem "is not really there to count *in the same way* that the other one is." One of the numbers tells *how many of the other one you have* "so it isn't really there" to count in the same way.

The adult multiplication explanations did not make as much sense to Wendy as her own explanation. Try to think like Wendy. In Wendy's mind, an addition problem like 6 + 2 meant that she had to count both six actual items and two actual items to show that she could combine them. For a multiplication problem, such as 3 × 4, Wendy had to count items for *only one of the numbers*. She had to "show one number amount as many times as the other number said," but as she saw it, *she had to count only one of the numbers*. Really listening to Wendy's explanation revealed that her thinking wasn't strange at all.

The placements of single digits in the place value columns require that those same digits are not there *in the same way as* when they are in different columns. Each digit in each column looks like the same digit as in all of the other columns, *but you do not count them in the same way* because the columns have been assigned different values. Children must comprehend those different values and, sometimes with a great deal of trust on their part, use that same-looking-digit in different columns. Sadly, far too many children learn to mechanically move around these digit placements without the understanding that Wendy had about their distinctions.

"TAKING APART" VERBS FOR SUBTRACTION AND DIVISION

The subtraction and division "taking apart" verbs are clustered separately in the two webs shown in figure 2.2 and describe the different verbs associated with these different and still very small inverse operation symbols. These graphic organizer webs also include more caution statements about how subtraction and division affect whole numbers differently than they do with those positive and negative whole numbers called integers and fractional numbers.

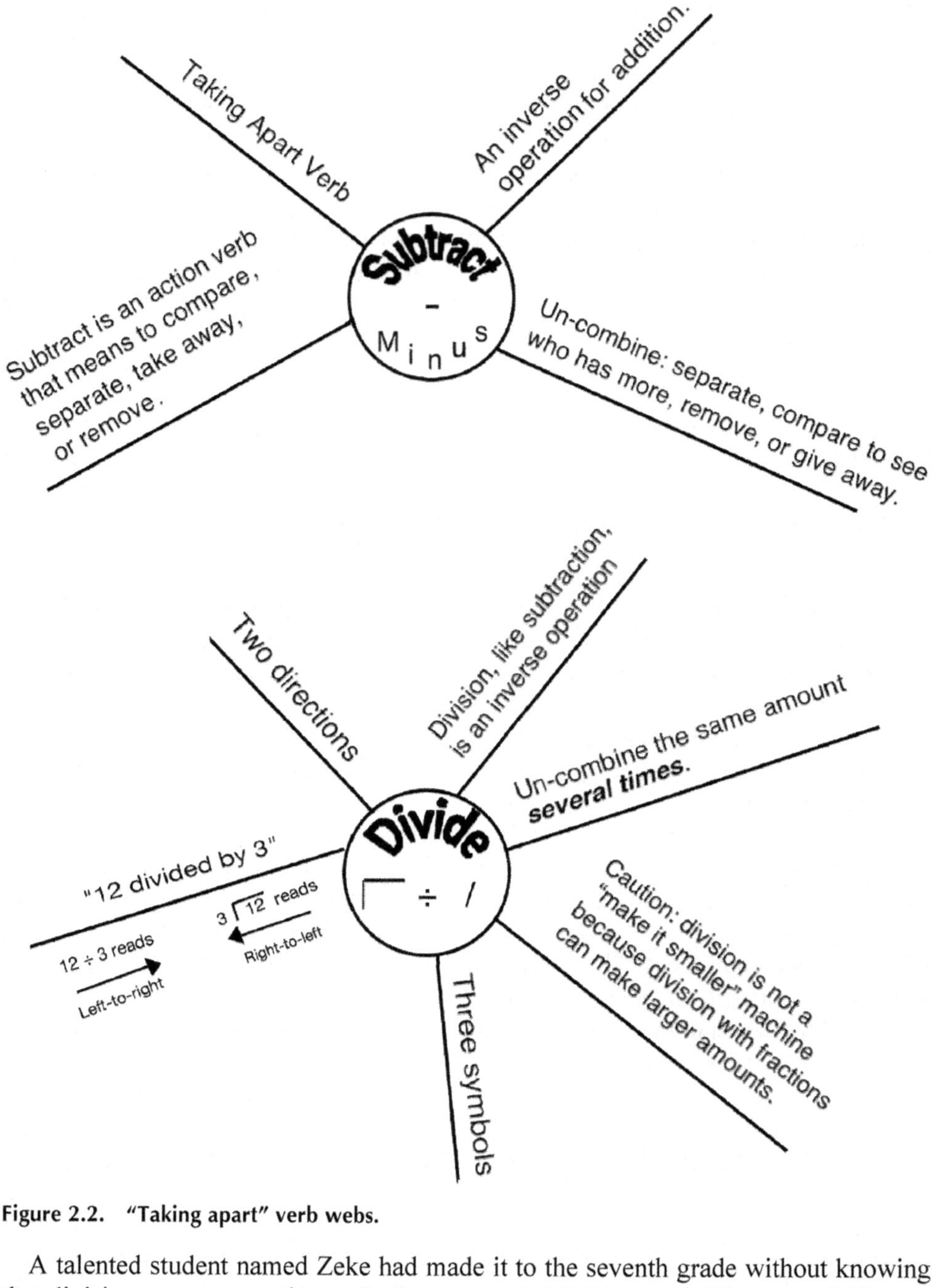

Figure 2.2. "Taking apart" verb webs.

A talented student named Zeke had made it to the seventh grade without knowing that division was a separation verb, the concept behind inverse of "put together" multiplication. He could "say it" but didn't know it as a verb action. When faced with long division without the calculator, he was constantly forgetting where to put everything in the long division "guzinta" format (a slang term for "goes into"). Be assured, Zeke is not alone. Many students enter the upper grades without knowing what division is really about. Zeke's problem was not short-term memory or working memory issues; the problem turned out to be mimicking the procedure.

Zeke was faced with the division problem, 168 divided by seven, and he wrote the expression as 168 ÷ 7 because the symbol and words matched the reading direction *and* this symbol was on his calculator. He had no idea which digits to place where, when to use them, or why they were used. Even the calculator was wrong some of the time with other division problems because the two symbols for division have opposite reading directions [(the guzinta symbol is read right-to-left and the 168 ÷ 7 is read left-to-right), sometimes 168 ÷7 and sometimes 7 ÷168], Zeke would press the calculator keys in an inconsistent order (a reminder that calculators don't think, they only respond to their keys). More on this "guzinta" expression is explained in chapter 5.

Instead of dividing on paper with a procedure or calculator, Zeke was asked to separate the 168 into seven circles, making sure that the same amount ended up in each circle. This task made sense to him so he started, without any help or glances for approval, decomposing the 168 into tens and other numbers. Eventually, he successfully separated the 168 by first decomposing the one hundred into ten tens, distributing the tens among the circles, and then he "broke down" (Zeke's words) the rest of the leftover numbers into more tens and fours (called number decomposition). His final rearrangement is shown in figure 2.3. It took time, but division finally made sense to Zeke.

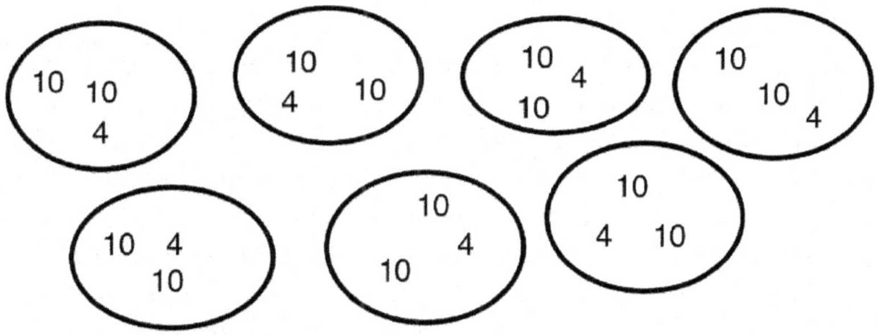

Figure 2.3. Zeke's division work.

Zeke's thinking:

Zeke decomposed the sixty into six tens because he ran out of the ten tens from one hundred before he finished filling the same amount in all seven circles. He used four of the six tens to finish the second round of distributing the tens. By now, Zeke had used fourteen tens out of the sixteen tens available from the 160. He was surprised and delighted to now know what guzinta (and division) meant, so that his only problem was how to "break apart the numbers to make them fit." Finally he decided to use the eight ones with the two tens. Now he had twenty-eight to separate. Fives didn't give him enough for the seven circles, so he used fours instead.

Zeke's real task in this division problem was to find the particular group of numbers that would be *in one of the seven circles*, the seven acting as the "dividing verb" number. By actually decomposing the number, he could reuse his thinking when confronted with different kinds of numbers, such as for dividing fractions, which are explained in chapter 3. Now that he understood the action involved with division, his next task was to translate his division into a format that resembled the one used in school. A more detailed explanation is presented in chapter 6 in a format that closely resembles the partial quotient method.

If division is only an assignment for your child to practice instead of a chance to work with "taking apart" verbs, then your child will have difficulty recognizing when to use division in word problems or any nontextbook situation. Divide means to split up, separate, share, or deal as long as the work is done equally. "The Greeks . . . found any problem involving division very much more difficult . . . they had no way of doing division to any order of accuracy."[3] Maybe the Greeks, many fourth graders, and Zeke have a lot in common!

A SHORT INTERLUDE ABOUT THE USE OF THAT NEGATIVE SIGN

Subtraction and division are the inverse "taking apart" operations for addition and multiplication. That "–" sign appears in different sizes and locations, yet each time it appears, an inverse is afoot. The early grades see the "–" as a subtraction between whole numbers. In later grades the "–" sign is used with negative integers. To make matters worse for the "location-challenged," that same "–" sign again reappears as an exponent and later in function notations. The good news is that the inverse concept itself doesn't change.

Subtraction does not seem to be a challenge until negative integers come into play, when a negative integer is subtracted from another negative number. This circumstance challenges everyone but the Fibonacci-originated debt industry.[4] That charge card notation for "debt" is noted as a negative amount on an accounting ledger. When the debt is paid, the debt is reversed, also known as a negative subtraction that brings the tally to zero debt. As part of an exponent (meaning multiplication), the "–" sign makes a number "take on an inverse implication." That "–" sign now turns numbers (ratios) "upside-down" (as students look at it) called a "reciprocal."

If this sounds like a mouthful, it is. The "–" sign always signals that an inverse is approaching. Two negative signs trigger two inverses. A warning about being cautious with the when, where, and why of these "–" signs is described more in chapter 8. For now, the Egyptians needed the reciprocal of a number so that they could divide. The Egyptians depended on successive multiplications called "duplation," so their version of division required the "invention of a number," hence the unit fraction that today is called the reciprocal of a whole number. The reciprocal is also called a "multiplicative inverse" and is further explained in chapter 8.

WORD PROBLEMS PUT MATH VERBS IN STORIES

Word problems tell stories and the verbs in the stories are the clues to choosing the operations for the calculations. Taking short cuts to solving word problems with sound bite clues—such as "and" means addition, "take away" means subtraction, or "difference" means subtraction—puts automatic limitations on all of the other possible options. Knowing about the verbs associated with the operations makes it easier for you and your child to interpret the stories described in those infamous word problems.

In the push to get those math facts memorized (supposedly because quick recall makes solving problems easier), you could forget to allow or encourage your child to solve problems on their own terms. The following solutions are by children who worked from their own way of thinking to solve a word problem about frogs.[5] For all of the children, the idea of frogs joining with other frogs clearly called for an addition setting, yet each child provided their own versions for achieving the same results. These children clearly have number sense and are comfortable composing and decomposing numbers.

Teacher: "Six frogs were sitting on lily pads. Eight more frogs joined them. How many frogs were there then?"

Rudy, Denise, Theo, and Sandra each answer "14" almost immediately.

Teacher: "How do you know there are 14?"

Rudy: "Because six and six is twelve, and two more is fourteen."

Denise: "Eight and eight is sixteen. But this is eight and six. That is two less, so it's fourteen."

Theo: "Well, I took one from the eight and gave it to the six. That made seven and seven, and that's fourteen."

Sandra: "Eight and two is ten, and four more is fourteen."

George Pólya earned the "Father of Problem-Solving" moniker because of his extensive work in developing thinking strategies for solving problems. He not only encouraged strategic guessing, but he also outlined four steps for solving word problems: understand the problem and see clearly what the problem story is about; look for the pieces that are connected and make a plan; patiently carry out the plan by choosing the operation(s) that you need; and look back to make sure the solution makes sense. Or, to include another option, look at the answer in the back of the book and then work backward to figure out why the answer makes sense!

When the "deer in the headlights" paralysis happens with the grand entrance of word problems, it could be that the first step even *before* Pólya's first step is an enigma. Perfection paralysis sets in when permission to guess and *not* be right the first time isn't an option. If your child thinks that good math means that you "must be right the first time," then your child could stay paralyzed by this threat. A byproduct of perfection-paralysis is a hope survival strategy—hope that the solution will miraculously appear, hope that the problem will disappear, or hope that the bell will ring. Paralysis inhibits thinking, and hope is not an effective strategy.

Another possible difficulty that your child might be facing is in the connection between the word problem story to the verbs associated with the operations. Since formulas are already finished expressions of relationships about patterns *developed by somebody else*, they may not help to understand *the actual story*. Using metaphors or analogies might serve to clarify the descriptions in the word problem. Sometimes just making up some similar stories with similar words provides clues. What *doesn't help* the visual-spatial learner (or any creative child) is to be told what to do with no imaginative contributions from them.

Listen to your child to learn about the conditions that may be causing their hesitation or paralysis. Avoid the temptation to "fix" your child's thinking. Wait for their thinking even if you need to decide on some time limits, like a ten-second window to make a choice. Then provide some hints that allow your child to "fix" their own thinking. Drawing a picture or a diagram *of their choosing* can also help. It does not matter what your child draws initially, as long as they use the problem context and break their paralysis. Something as seemingly inconsequential as drawing a simple picture can provide the aspirin to relieve word problem headaches.

Encourage your child to choose one word in the word problem so that they have *a single focus* for what to draw. If they are too timid, then you choose one. For example, if the problem story describes framing a picture, then choose "frame" and let them draw a frame. If it is a designer frame that is different from the usual rectangular frame, then the story problem will tell you so, and your child can draw that designer frame. In most word problems, a picture can help the visual-spatial learner understand what the story in the problem is really about.

Clues come from the phrases in the story. For example, frames and fences could suggest that a perimeter is needed, and perimeters usually suggest addition. If the storyline includes area, then multiplication is likely involved. Pictures like the ones in figure 2.4 can help your child get started. The rest of the word problem story will describe a relationship that involves either the frame or the fence. Either some dimen-

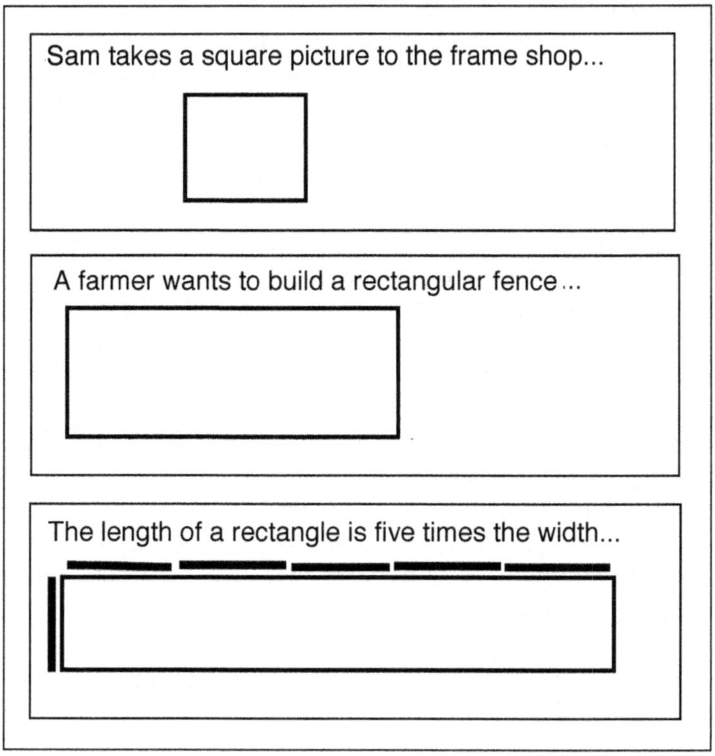

Figure 2.4. Drawing examples for pictures, fences, and rectangles.

sions of the frame, or the length or area of the fence, will be mentioned, such as when a word problem describes a relationship between length and width in a rectangle with a phrase like "the length of a rectangle is five times the width."

Many times one picture can serve for several different kinds of questions, just by adjusting the measures in the picture. One possibility for a fencing question could be, "If the yard is eight feet by three feet, how much fencing is needed to go all the way around the yard?" Or a different question could be, "If the fenced in yard is eight feet long and you have thirty feet of fencing material, how wide could the fenced-in part of the yard be?" The same picture can apply to both problems with the length measure replaced with a question mark as shown in figure 2.5.

A framing problem could be, "If the picture is three feet square, then how much frame material is needed?" If your child asks about how much to add for mitered cuts, then you know that they have a good idea of the storyline as well as woodworking. A different question for the square picture frame in figure 2.5 could be, "If you have twenty feet of framing material, do you have enough to frame a three-foot square picture?" Does 3 + 3 + 3 + 3 total more than twenty feet (even with the extra length that it takes for mitered corners)?

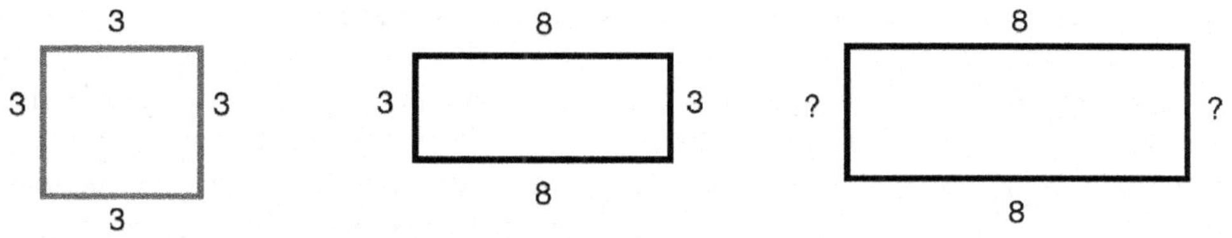

Figure 2.5. Different questions about framing a picture.

Selecting one phrase for a possible picture can suggest which operation to use, such as the examples in figure 2.6. A picture can be a lead-in to identify what information is needed to respond to the question. Again, one picture can have multiple scenarios. If the word problem describes a relative length between people's heights or relationship between people's ages, then one operation might work just as well as another. Using the same "times three" relationship that the father (or mother) is three times taller (or older) than the son, the questions could be, "If the man is six feet tall, how tall is his son?" or "If the son is ten years old, how old is the mother?"

As to which operation is best, that decision is best made by the problem solver. One child might choose addition while another child could choose multiplication. Both involve the "putting together" operations. Some say multiplication is more efficient and should be the operation of choice. However, efficiency too soon is just as damaging and destructive to thinking as perfection-paralysis. Efficiency can come later but *never* at the expense of making sense.

Either addition or multiplication could solve the story about the heights of a man and his son. The operation that bests solves the problem depends on what the question is asking and the what operation makes the most sense to the problem solver. Different kinds of questions would suggest different operations. Since all four of the operations are related, then your child might choose a combination of operations. Which operation, or operations, your child chooses is not as important as the fit to the question and the accuracy of the results.

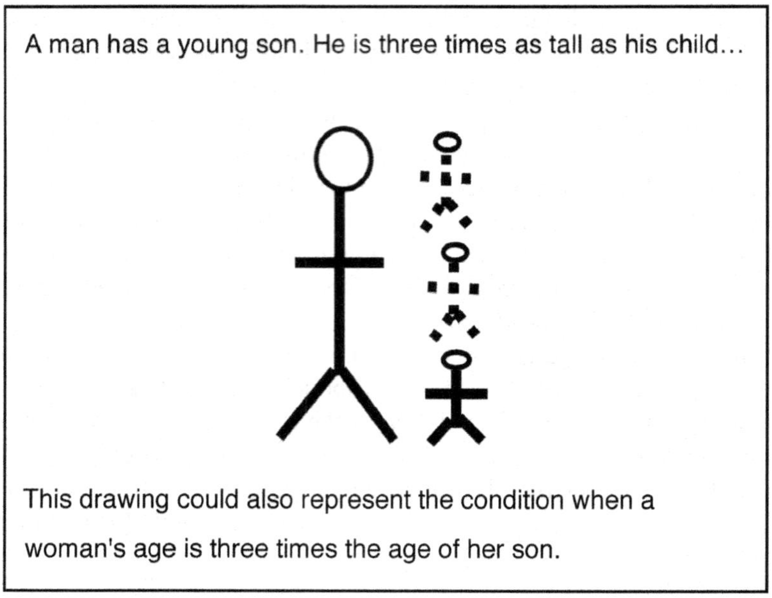

Figure 2.6. Three times age or three times taller, same picture.

A fifth-grade teacher used cartoons from the newspaper to help her children write a storyline to develop language skills and also to demystify word-problem writing. Her students cut out separate comic strips from a newspaper, separated the pictures into frames, and put their separated frames into separate bowls. Using cut-out context pictures can encourage your child's creativity by allowing them to give names to the characters, contexts to the events, and descriptions of each character's personality.

Finding separately framed pictures from comic pages or even in some magazines can encourage children to change the order sequences and, therefore, encourage different stories. A third grader named Leslie made up her own story, drew her own pictures, and then wrote her own math story. The only requirements were that she had to choose at least one verb and one question to include in her story. The picture in figure 2.7 represents her story, which involved two people each having 3½ cakes and 9 ice cream cones *with* jimmies. Leslie's cakes cost $4 each (half cake $2). The cones cost $7 each. Leslie's story was far more involved than the ones in her book.

Then there is John Paulos' suggestion or Enrico Fermi's approach to engage students' imaginations with questions like, "How many watermelons would fit inside of the White House?" or, "How many basketballs will fit inside Boston's TD Garden?" and, "How many elephants will fit inside this room?" A third-grade student named Alice immediately replied to the elephant question with, "Which country is the elephant from?" She was too young to know about those elephant size differences, or was she?

Questions like these encourage a rough guesstimate and a search for information using any available resources. Even though your child's guesstimate may seem outlandish to you, these kinds of problems are still curiously tantalizing. Outlandish is okay! Tantalizing is even better! By using these and other similar questions, you and your child will be engaging the most important resource, your child's imagination.

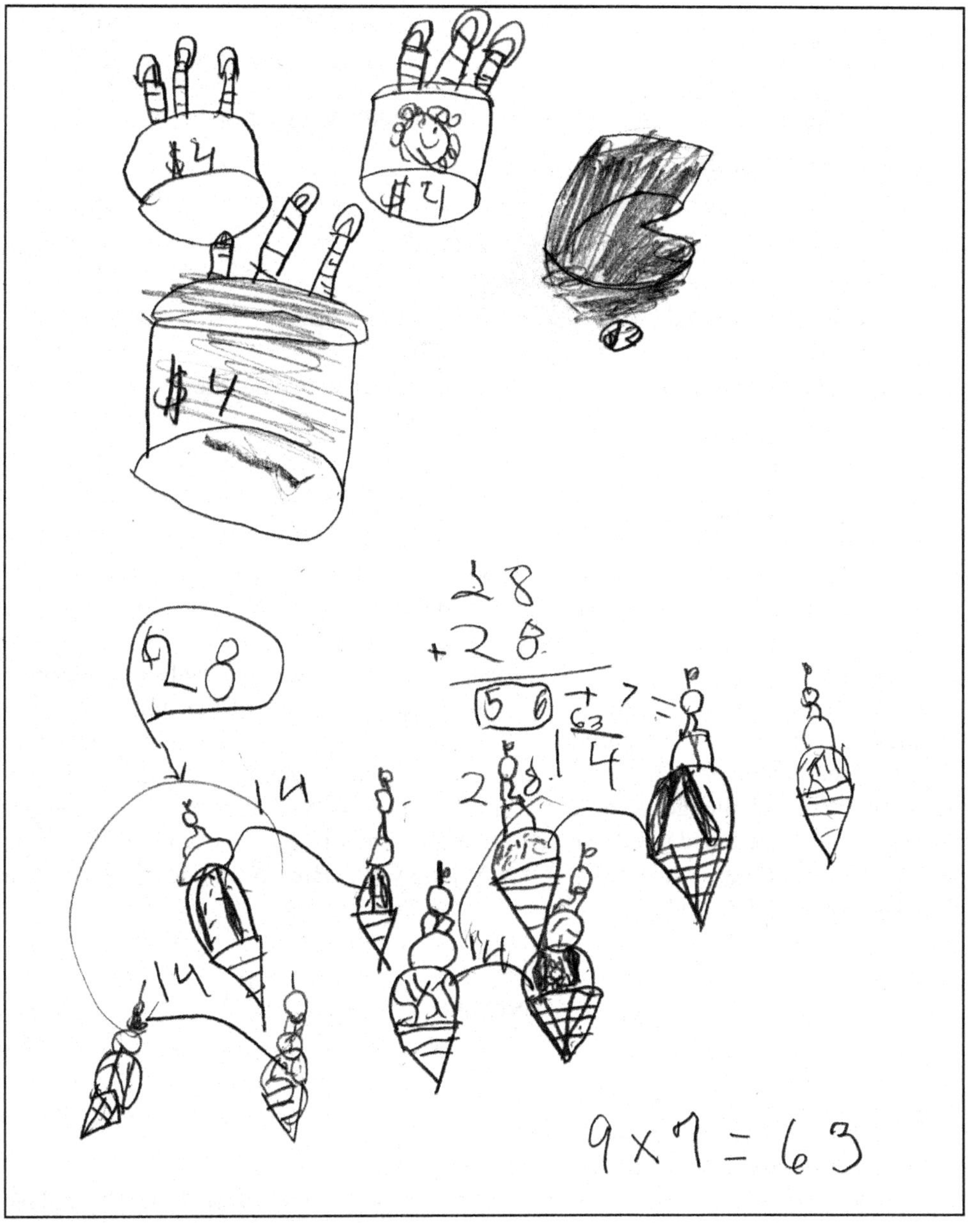

Figure 2.7. Leslie's picture for her story.

Try This:

Figure out how many pizzas are consumed each year in the United States.

What information do you need to know?

What information would you look up on the Web?

Where else could you find information to help calculate the number of pizzas?

Figure 2.8. Try This.

KEEP IN MIND

The symbols for operations (+, –, ×, ÷) may look very similar to each other through your child's eyes, so take time to "translate" them. Be wary of the imitative mathematics or "fake math" or "sterile math" as you use these symbols with your child. Imitation mathematics requires your child to just "do the math" without asking any questions. Your challenge is to help your child be involved and understand the appropriate and understandable language for these symbols. Help your child feel good about "figuring things out," even if it is the "long way." Brevity should never come at the expense of understanding.

NOTES

1. Davis and Hersh, *The Mathematical Experience*, 140.
2. Boyer, *A History of Mathematics*, 15.
3. Hogben, *Mathematics for the Millions*, 21.
4. Eves, *An Introduction to the History of Mathematics* (, 233.
5. Peterson, Fennema, and Carpenter, "Using Knowledge of How Students Think about Mathematics," 24.

Chapter Three

The Choice between Two R's

Rote Memorization or Reasoning

> Freed from the tyranny of memorizing, all of the energies of the pupil are bent on thinking and what is once clearly thought through is already half remembered.[1]
>
> —J. W. A. Young

Memorization is a choice; it is *always* a choice. Choosing between memorizing and "thinking through" to make sense of math does not always seem like a choice, but it is. The myth is that memorizing works for everyone. Memorizing may work for the few who have "memorizing minds" (or are auditory-sequential learners),[2] but many visual learners do not wish to use their hundred billion neurons that way. Some want an instruction manual, and others, like Jen, Wendy, Bobby, Trevor, Jim, and maybe your child, too, want to figure it out on their own terms; they choose to *think through the information* and *then* create their own version.

This chapter describes a few examples of how memorized slogans, also called sound bites, can actually interfere with learning, and as a result can lower test scores. Thinking through the craft of mathematics is more dependable in more circumstances than those sometimes successful, rote memorization survival slogans. The craft of mathematics is not about "two negatives make a positive" or "moving decimals" or any sound bite tricks or shortcut mystical rules that get answers, *sometimes even right answers*. The backbone of *real* mathematics *is* about how all of these numerical expressions are interrelated with relationships and patterns.

Parents only want success for their children. They face the math-learning conundrum, "To memorize or not to memorize," when they themselves may not understand the math. The noble advocates on either side of the argument have always found sufficient support for their respective positions; however, research shows that rote memorization, drill, and practice does not work for the visual-spatial learner.[3] The discussions about the benefits of "drill and kill" math practice (usually using rote memorization) over building meaningful learning in mathematics (usually using relationships) will no doubt continue for decades to come.

These debates tend to be fueled by each new innovation or new initiative that usually follows from poor results on one of the many standardized tests. The debate is

	Percentage Recall		
	Immediate	After one day	After 4 weeks
schematic	69%	69%	58%
rote	32%	23%	8%

One group was "taught" by schematic instruction that included an inherent scheme or system such as pattern recognition and relationship recognition. The other group was instructed by a rote memorizing instruction where children were told to memorize the rules.

Figure 3.1. Skemp's research results.

far less important than the child's understanding of this thing called mathematics. Besides, when freed from the tyranny of memorizing, your child's energies are spent on *thinking and remembering*. Researchers are still measuring the pros and cons of memorizing versus the thinking methodologies. One of the early research results[4] is shown in figure 3.1. It seems far more reasonable to bank on the 58 percent successful recall strategies after just four weeks.

If rote memorization has only an 8 percent success rate after four weeks, then it would make more sense to make the effort to increase the success rate to 58 percent recall. As long as you and your child are putting in the time to learn math, then it makes more sense to use the relationships and patterns that support recall and that get results that last longer than a month. Shortcuts and sound bites, while well intentioned, can do more harm than good. They work only *some of the time* and tend to put another confusing layer onto something that is already fairly natural. Besides, when math is clearly thought through, half of the work of remembering is already done.

SOUND BITES THAT BITE BACK

Do any of these sound bites sound familiar? "The larger number always goes on top." "Can't take a larger number from a smaller number." "Multiplication makes bigger numbers." "Subtraction makes a smaller number." "Division makes smaller numbers." "Always distribute the number outside the parentheses to everything inside the parenthesis." "Two negatives make a positive." "Invert and multiply (or keep, change, fix)." Each of these sound bites satisfies one condition *once*, but there are *several* circumstances in which *they do not work*. Sound bites can cause lower test scores because of all those other times *when the sound bites do not work*.

The tricky piece is that sound bites have a vise-grip survivalist hold on that *one time* when the bite actually works. These one-time helpful sound bites eventually need to be unruled and unraveled, thus confirming math's reputation of being mystical, secret, magical, and confounding because the "rules" are always changing. Math may not be easy, but it certainly does not deserve the mystical, magical, and confounding status that it has to the Math Avoider population. In real mathematics, as opposed to "mimicry math,"[5] the rules do not change. Sound bites are the "wrong rules."

Sound bites for whole numbers do indeed bite back when fractions come into play. One of the most notorious offenders is: "The larger number always goes on top" (how about those regular fractions that are not improper fractions?). The fractions that children usually use are the proper kind. Another is, "Multiplication makes bigger numbers"—unless you are multiplying two fractions, in which case one of the fractions gets split up into even fewer pieces. What about "Division makes smaller numbers?" Not so with fractions or negative integers. Division is used to find out "what's in the group," so different kinds of numbers yield different group amounts.

"Always distribute the number outside the parentheses to everything inside the parenthesis" does not apply to exponents. When the number "outside of the parentheses" is a coefficient (the number in front), then distribution is appropriate. However, if the number "outside of parentheses" is an exponent, then the exponent is not distributed *because it describes a different kind of multiplication.* Those exponents have an association with place value and the multiplication relationship among the columns, while coefficients have an association with addition. See more about these differences in the "Acronym" section below.

The statements that you "cannot take a larger number from a smaller number" and that "two negatives make a positive" are integer (positive and negative whole numbers) downfalls. You "cannot take a larger number from a smaller number" is often repeated as a guide for young learners to learn how to write those first subtraction problems. For some reason, this sound bite has a very long and tenacious math life and bites back as soon as integers are introduced. Countless wails and "but they said" blame-filled choruses fill the air and invoke raised and frustrated voices around the homework table and in upper-level classrooms.

"Two negatives make a positive" dumbfounds many because this sound bite *only applies once* when multiplying negative integers. Ruth McNeil expresses hers and others' lament when she writes, "What did me in was the idea that a negative number times a negative number comes out to a positive number . . . underneath, however, a kind of resentment and betrayal lurked, and I was not surprised or dismayed by further foolishness my math teachers had up their sleeves."[6] How negatives can magically "change coats" to become positive has the mystical and magical aura of a secret code, but, in actuality, has a reasonable explanation (see figure 3.2).

Round disks of two different colors can provide a visual representation for positive and negative integers. The white color represents positive integers and the black color represents negative integers. For example, to represent the integer problem "+4 + –7," the +4 chips are all white and the –7 chips are black. To combine these two colors (add the integers), match every white disk to a black disk and then count the leftover unmatched disks. The same disks can be used with multiplying integers; just make sure that the discs are put into groups for the multiplication. Figure 3.2 shows the integers used with both addition and multiplication explanations.

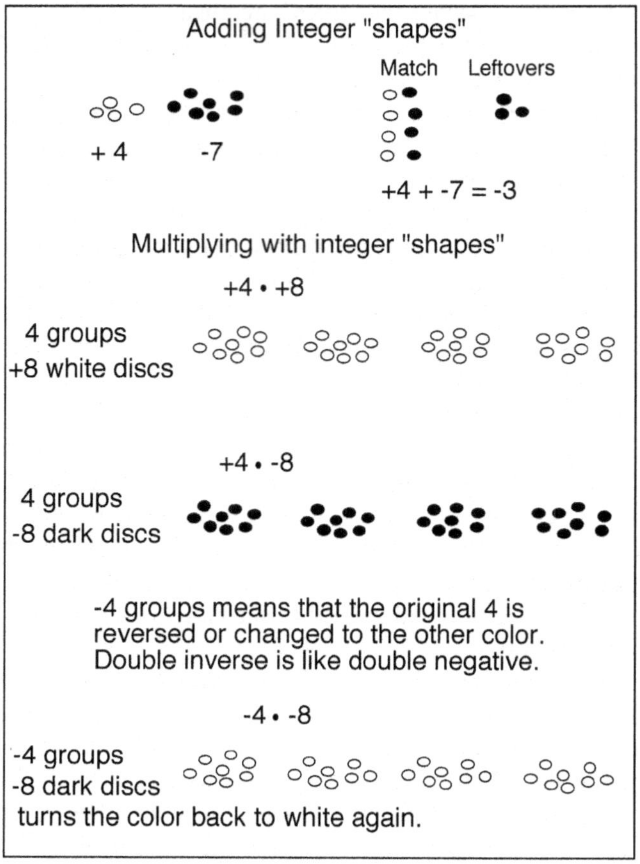

Figure 3.2. Adding and multiplying integers.

Remember, only black and white or positive and negative are available for these integer operations. One inverse turns the white discs into black discs. The second inverse makes them white again. To determine the color of the resulting discs, keep track of the inverse changes of color. It is no more complicated than keeping up with flipping discs and the related number of inverse uses. We will explore more examples in chapter 7. Do you wonder if this black and white interpretation would have helped McNeils's feeling of betrayal about math foolishness?

"Invert and multiply" brings with it yet another version of "fake math," as the often quoted Steven Leinwand comment says, "Yours is not to reason why, just invert and multiply."[7] No thinking or reasoning about why, "just do it." This sound bite is another "quick fix" in the long list that peppers what is generally called "imitation math." The "invert and multiply" mantra creates a knee-jerk response in far too many children. When they see two fractions or ratios, their knee-jerk reaction is to invert and multiply or cross-multiply, whether or not there is a "÷" or an "=" in the expression.

The verb "invert" suggests that a reciprocal is involved as a multiplicative inverse from chapter 2. That reciprocal form sounds more than a little ominous and complex than it needs to. It first came into being so that the Egyptians could divide, and we

Try This:

Find some black and white counting chips or counting bears - or any two colors of the same kind of object.

Combine the items. Think about how colors will represent the final amount.

Try these multiplication problems:

$-4 \cdot 6$	$2 \cdot -5$
$3 \cdot -7$	$5 \cdot -5$
$2 \cdot 5$	$-8 \cdot 3$

Make up some more multiplication problems to try. Keep a record of the numbers that you try and the results.

Remember Wendy's declaration about multiplication that one of the numbers is not there in the same way that the other one is.

You will need to select which number "is there" and which is "not there." Try the numbers both ways.

Figure 3.3. Try This.

still use it today to signal that "an inverse is in the house." The "and multiply" expression continues with a conversion from division to multiplication, representing *another* inverse application. It is very much like using a double negative in a sentence. Double negatives in sentences simply double "undo." They, too, can put a layer of confusion into a simple sentence.

MEMORIZATION VS. ORGANIZATION

Pete was totally focused on "facts" that he had memorized quite well. After all, the numbers always behave the same, right? Not always. Not understanding the organization idea of the place value columns means that your child and Pete have only part of the "calculation picture." Accurate calculations require more, much more, than simply knowing "facts." Without the organization idea, calculating feels like putting together a jigsaw puzzle without all of the pieces or without the cover picture of the box, or at least that is what some children say.

Prove It to Yourself:

The "rule" says to invert and multiply. Which ones and why?

$$12 \div \frac{1}{2} = 12 \times 2 = 24$$

Division is an inverse operation and noted with the ÷ symbol. Fractions also means there is division happening. Which one are you supposed to invert and why? The answer is, both of them. The inverse symbols need to be returned to the non-inverse status because it really means reverse the reverse.

$$12 \div 1/2$$

You are faced with working with two inverse operations (divide and reciprocal). Cut twelve boxes into halves.

You have 12 squares that are divided into ½ pieces.

How many half square pieces?

There are 24 half squares that make the original 12 squares.

Figure 3.4. Prove It to Yourself.

The examples in figure 3.5 are from one of Pete's papers. Pete's strategy worked sometimes, but not always. How confusing is that? Pete's homework included all kinds of two-digit addition problems, some that required regrouping and some that did not. Pete had all those basic addition and subtraction facts down pat, yet they were not reliable. Pete was well on his way to becoming a Math Avoider because his best and

36	25	78	95
+ 78	+ 10	- 36	- 18
1014	35	42	83

Figure 3.5. Pete's Work.

earnest "basic fact" strategies were not dependable. Pete used the same strategy for all the problems, sometimes the strategy worked and sometimes it did not. Pete started believing that he did not have that "secret code" for math.

Even and odd numbers can be represented with rectangles similar to the ones developed by some fourth graders as part of a number line activity about factors. The fourth graders called these number arrangements "spatial number personalities"[8] shown in figure 3.6. Organizing these spatial number personalities shows how they fit together. The even numbers are rectangles with a "two" for a side measure; the odd numbers (not rectangles) can also have a "two" for a side measure, *but with an extra square* in their area. Thinking like Daniel Tammet, a noted autistic savant with numbers, means that making even numbers is the same as creating pairs, so that pairs can even out odd numbers and even number pairs make more even numbers.[9] Tammet says, "If even numbers have half parts, their sum will have a half part."[10]

Ask Yourself:

Both even and odd number spatial personalities have a two on one of the rectangle sides. What do they look like when added?

Even number example:

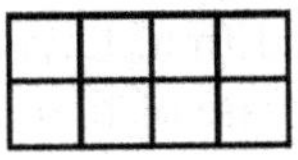

Odd number example:

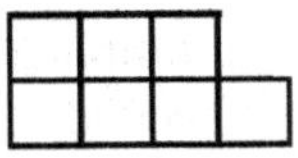

Cut out some even numbers and some odd numbers as rectangle spatial personalities.

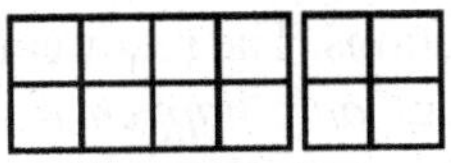

Even plus even

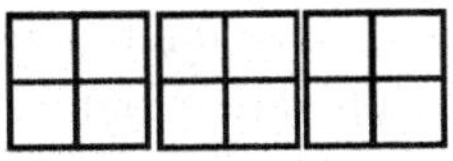

Even plus even plus even

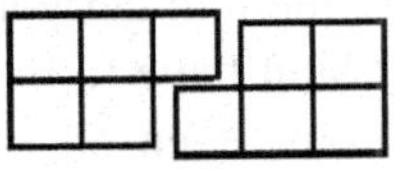

Odd plus odd

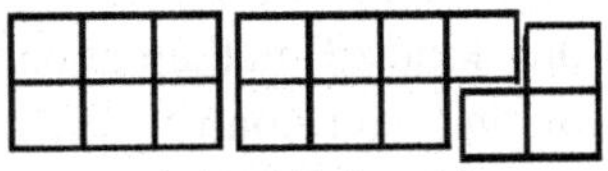

Even plus odd plus odd

Add the spatial personalities to see what they look like when they are combined.

Will odd numbers combine to make an even number arrangement or will they generate another odd number?

Figure 3.6. Ask Yourself.

ACRONYMS: SENSE OR NONSENSE

There's nothing wrong with acronyms, unless there *is* something wrong. An acronym is supposed to help remember the "steps" using a mnemonic for guidance, but when it bears no resemblance to the actual mathematics craft then it can confuse more than help. The "All Cats Race Only New York Mice" acronym phrase is an example that has no meaning other than the letters of the "acronym." "Please Excuse My Dear Aunt Sally" or PEMDAS is another often used example for making decisions about operations, and "Does McDonald's Sell Burgers" is the sequence for divide, multiply, subtract, and bring down in long division.

The McDonald's mnemonic includes "bringing down" a number as part of the process even though "division has nothing to do with directions such as up and down."[11] This use of nonrelevant information as part of an already confusing procedure encourages children to *not think* and *just do*. If your child is "just doing" the long division or other math procedures for survival rather than making sense of the craft, then, at best, they are practicing mimicry and "imitation math." At worst, this is one of the junctures for totally giving up on math reasoning and understanding!

Did anyone explain why the order of operations is even required, other than to say that two different calculation "answers" are not acceptable for the same problem? Actually, that *is* the reason that the order of operations is required, but *why is PEMDAS required?* The order of operations is about grouping the numbers *first* to guarantee the same results. All of the operations are extensions of addition, so it would make sense for the multiple addition (multiplication) to be completed first. Division is related to multiplication, so it comes second. Get the groups taken care of first and then "follow the signs" for the operations.

The first decision about groupings in PEMDAS is between the *Please* parentheses and the use of any *Excuse* exponents. Numbers in parentheses are dealt with first since that is the first grouping (unless the algebra class is using parentheses with two terms that are not combinable, such as "x" and "y"). Just the existence of parentheses signals that a *grouping* is in effect, not a requirement for operations. The requirement depends on the numbers and the operation sign. The operations *only happen* if the numbers allow it. Exponents represent a *specialized multiplication* so that must follow in the next order of priority. The arrows in figure 3.7 are there to get you started.

The groups in any kind of mathematical expression have to be taken care of first. Control the groups and you control the calculations, even on a calculator. If your child's calculator is not programed to recognize *PEMDAS*, then they must do the *thinking first* and *then* press the calculator keys in the correct order. Mathematics requires accuracy and consistency of results, even from calculators.

KEEP IN MIND

There are no effective long-term shortcuts to the operations, primarily because the operations are simple enough as actions—add means combine, multiply means combine *in groups*, subtract means remove, and divide means *remove in groups*. The

Prove It to Yourself:

Apply the PEMDAS order on these expressions.

$$(8 \cdot 5)/4 + 6(3 + 5)^2 - (5 \cdot 6)/10$$

Please: 40 and 8 and 30

Excuse 8^2

My: $6 \cdot 8^2$

Dear: 40/4 and 30/10

Aunt **S**ally: $10 + 384 - 3 = 391$

Explain what would be wrong with adding the 4 + 6 first.

Figure 3.7. Prove It to Yourself.

inverse operations simply "undo" the original put-together operations. The antidote to memory loss on those infamous tests is found in the effort to clearly think through the information so that it makes sense. As one student shared, "I learned that sometimes rules can trick you. So be very careful."

NOTES

1. Stein, "Young's Vision," 331.
2. Silverman, "Identifying Visual-Spatial and Auditory-Sequential Learners," 23.
3. Silverman, "Identifying Visual-Spatial and Auditory-Sequential Learners," 23.
4. Skemp, *The Psychology of Learning Mathematics*, 26.
5. *Everybody Counts*, 70.
6. McNeil, "A Reflection on When I Loved Math and How I Stopped," 45.
7. Curcio, "Divide and Conquer, Unit Strips to the Rescue," 6.
8. Fitzgerald, "A Number Line with Character," 368.
9. Tammet, *Thinking in Numbers*, 49.
10. Tammet, *Thinking in Numbers*, 69.
11. Kamii, "Arithmetic: Children's Thinking or Their Writing of Correct Answers," 2.

Part II

ORGANIZATION

> Never tell people how to do things. Tell them what to do [what you want] and they will surprise you with their ingenuity.[1]
>
> —General George S. Patton

Children's ingenuity in mathematics can compare, or even rival, the thinking of the early mathematicians. Both today's children and the many mathematicians that came before them have to keep faithful to the place value that organizes the operations, a definite equalizer of mathematical combinations. The chapters in this "Organization" section are all about how that place value organization plays a crucial role in the various methods used for math operations, whether they are "new math" methods that your child faces now or the "old way" that you might have learned.

Almost every decade brings with it new methods for calculating math, groundbreaking ideas, or new evaluation legislation. These events can create an atmosphere of discontent with the current math programs. This dissatisfaction brings with it new descriptions about how to add and multiply, or, heaven forbid, change long division. These new methods suffer the ire of frustrated parents declaring the new descriptions as "new math," regardless of how improved those calculation methods or teaching guidelines may or may not be. What never changes is the craft of mathematics and number sense and the organization *behind* any of the "new" methods.

Chapter 4 describes the importance of classification and organization to the operations of adding, subtracting, multiplying, and dividing. These operations combine and take apart numbers, but the numbers need to *belong to the same category* in order for the combinations to make any sense. Chapter 5 shares children's stories about their interpretations of the operations. Chapter 6 looks at the "old way" procedure and compares that outline with several "new ways" to add, subtract, multiply, and divide. The key is to compare, not to learn a new method. No single procedure is best for everyone.

With calculators so readily available, it is possible to forget how confusing it can be to learn about the procedures involved in the operations, or even to be patient about your child's need to learn them. Or, as many parents have confessed, you *can remember* with vivid detail that pivotal moment when you gave up. Having that number sense idea of composing new numbers by adding, subtracting, multiplying, or dividing with other numbers is vital to your child's understanding of how those operations affect the rest of mathematics. Having flexibility with how new numbers are generated is the best strategy for success.

Actually, some of the "new ways" are not new at all; some date back centuries and some are very similar to what today's children have intuitively done on their own without "formal instruction." The only differences seem to be in organization and notational preferences. Some cross out, some do not. Some erase, some do not. These preferences are centuries old and evolved over time as necessity dictated, from writing on tablets to the invention of paper and on to computers that have very large virtual "pages."

Your child can even make up their own procedures using number sense and organization, as did the children featured in these chapters and the mathematicians of "olde." The idea is to help the Math Avoiders-in-the-making slow down and not fall prey to Robbie's reasoning. One student, Robbie, reasoned: "It is better to go ahead and do it fast even if it is wrong because you get in less trouble that way." Or, he may have gotten the right answer, but it wasn't the "right" way, so he didn't get credit for his work. There are many ways to accomplish accurate operations. It is best to *not* fall in love with any particular method.

NOTE

1. https://www.brainyquote.com/quotes/quotes/g/georgespa106027.html

Chapter Four

In Search of Like Terms

Classification Revisited

If you focus on just the answer, then you will miss the mathematics.[1]

—George Pólya

A refrain often heard around the homework table is "you cannot add apples to oranges." An apple is not an orange any more than a goat is a chicken. However, when there is an agreement for common category, oranges can be added to apples. Math requires the same common agreement before any combinations are possible. That common agreement is called "like terms" in math speak. The common category for apples and oranges is "fruit" and a common category for goats and chickens could be "animals." This chapter explains the "apples and oranges" of like terms and how this concept makes those "rules" make more sense.

LIKE TERMS, COMMON DENOMINATORS, AND SAME UNITS

Keeping like terms together makes combinations possible, whether you use addition, subtraction, multiplication, or division. Keeping track of like terms as necessary in order to keep track of the results is another one of those major ideas in mathematics. The words "like terms" are not usually used until algebra or pre-algebra, yet the concept is used as early as first grade with apples and oranges and later with adding fractions. For those who have understood the significance of like terms, the restriction is reasonable. For the rest, if nobody warns them when it is okay to use like terms and when it is not, the restriction can seem whimsical.

Don't wait until your child is in algebra to talk about the "why" reasons behind the real mathematics of using like terms, like units, and like categories. The key to the common agreement is in the common category for the items. In measurement, inches are added to inches, but inches cannot be added to square inches because one is a line and the other is a square. Inches can be added to feet after feet measures are recategorized (or converted) to inches. Pounds cannot be combined with grams unless there is a common category (or another conversion) so that both are expressions in the same (like) units.

Measurement units, fractions, ratios, and algebraic letters such as x's and y's require identifying common categories before combining the units. The three parts of figure 4.1 show three conditions and related examples: "Like Terms," when items can be combined; "New Category," when items need to be converted to the same category before combining; and "Unlike Terms," when the items cannot share a common category so cannot be combined. Your child can combine numbers of items that may initially appear to be different *as long as a common category exists that fits with both numbers*.

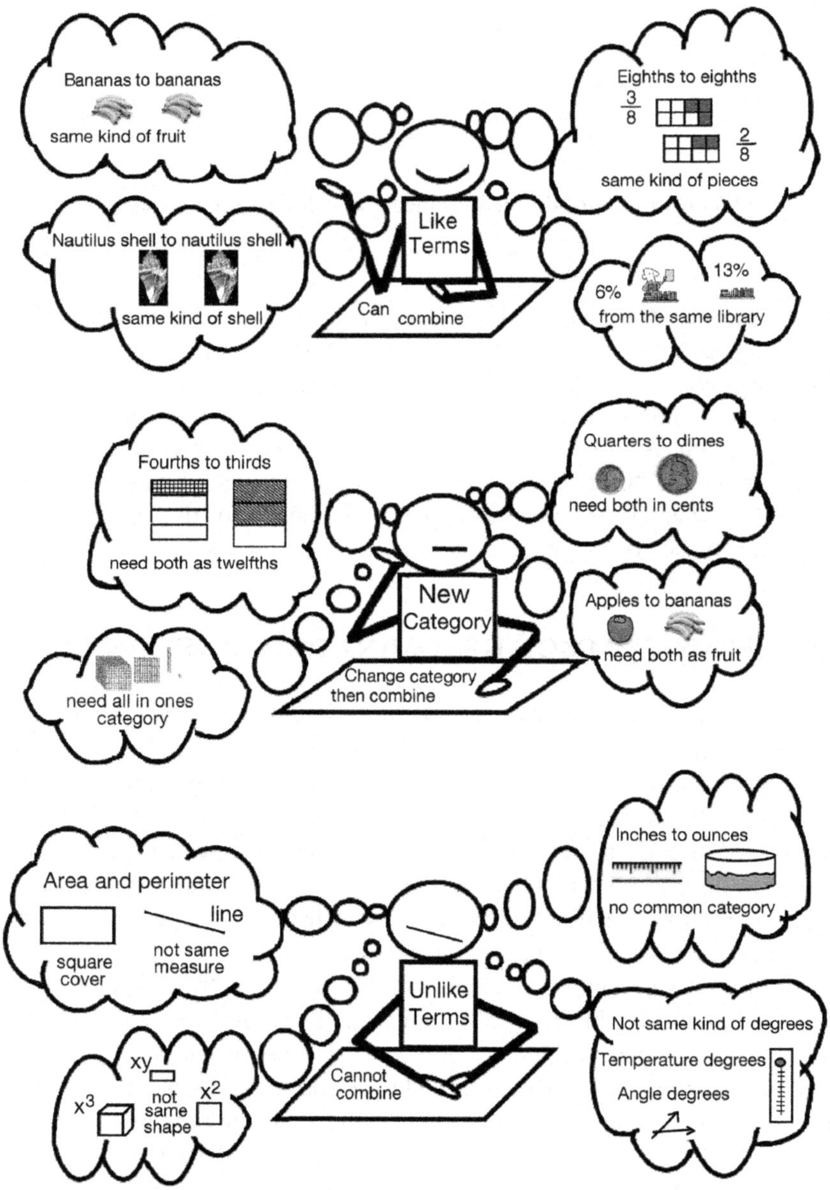

Figure 4.1. Like terms, new category, and unlike terms decisions.

Your child can now reasonably express the combinations of hundreds to tens and also to ones in a single number amount because all of these blocks can be fit into a category called "ones." Hours can be combined with minutes because both hours and minutes can be converted to minutes. Coins, like quarters and dimes, can be added because they both share the category of cents. Numbers for bananas and apples can be combined as long as the total represents a total amount in a fruit category. Lets look at more circumstances when we need common categories.

PLACE VALUE COLUMNS HAVE LIKE TERM CATEGORIES

Adding like terms appears *within* each of the place value column categories. Your child adds or subtracts hundreds with hundreds, tens with tens, and ones with ones. The same column categories apply to subtraction. The number decompositions needed in order to break the categories in the subtraction process (very much like "breaking a dollar to get change for the parking meter") has had several names: "carrying and borrowing" to the 1950s crowd, "trading" to the 1960s crowd, "regrouping" to the 1980s crowd, and composing or decomposing, or rearranging for the children of the new millennium.

Descriptions about when you can and when you cannot combine hundreds to tens and to ones may sound contradictory. Without an understanding of the difference between a "procedure" and just saying a number, they *are* contradictory. A procedure is an algorithm (that is math-speak for how to do the calculation) and is set up to keep the column categories *separated.* Subtraction and division are the two operations that can require some decompositions to "break into" adjacent columns. When a number is just said out loud, then the number expression is already sectioned into place value categories in order to say it.

Keeping the common categories separate is why the place value organization actually works—that place value structure organizes *all of the ones* that are in every given number. Sometimes that message can get lost when there's so much focus on the procedure and *where to put the digit,* not *why it goes there.* When your child understands that it isn't as much about the columns and digit placement as it is about the organization category for ones, then a reasonable math world opens up for them.

Figure 4.2 shows a typical traditional layout for adding two four-digit numbers, but with some adjustments that will hopefully provide some clarity around the "why" of digit placements. This layout includes "zone areas" that indicate the category titles, translations, and conversion areas. The zones show the explanations. Each vertical column has four different ways to describe the title and all carry the same message: a word title, an algebraic title, a number title, and a shape title. The conversion zone is where the *reassembly* happens either after (or during) completing the "work of the problem."

Only the same shapes (categories) can be combined in this organization—thousands large cubes to thousands, hundreds flat blocks to hundreds, tens sticks of tens to tens, and ones to ones. Figure 4.2 shows the two groupings of blocks that are represented in a traditional addition problem of 1,475 + 6,523. The shapes for the organization are

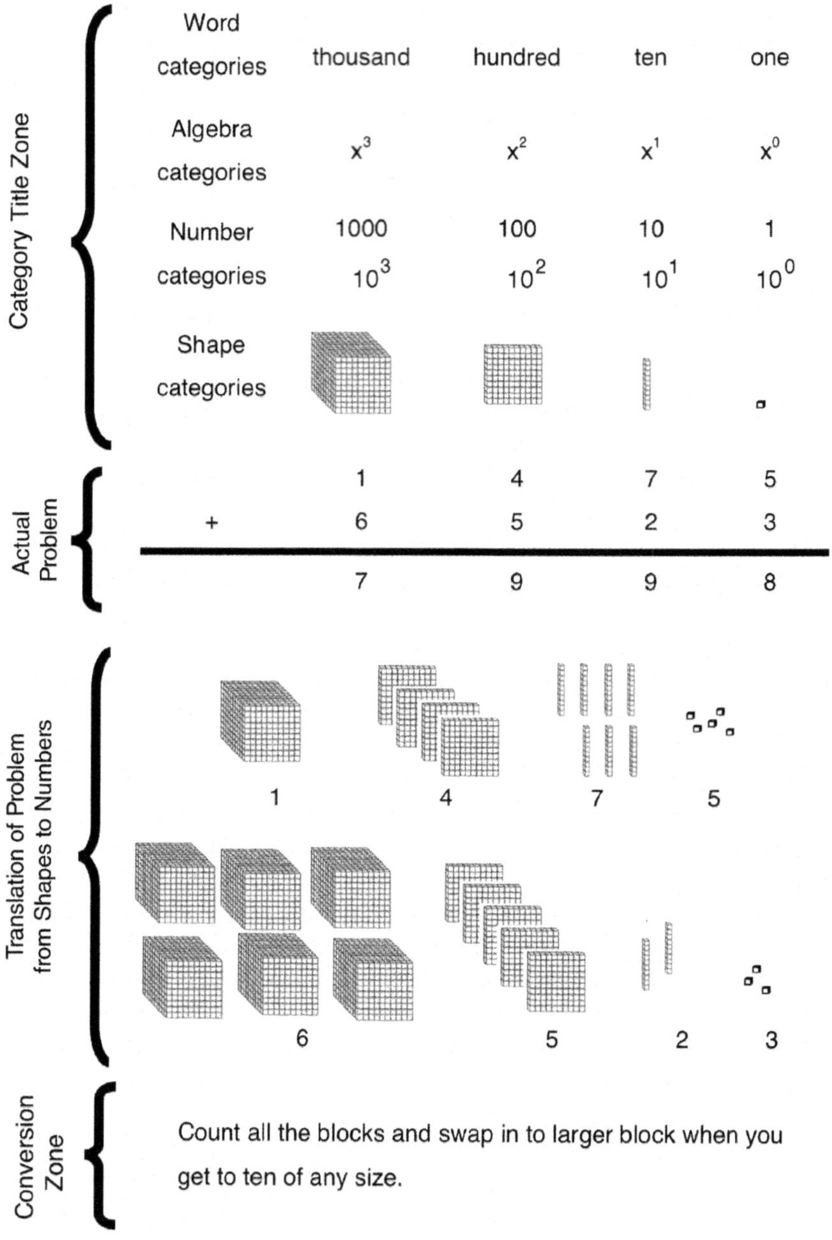

Figure 4.2. Format zones.

the labels at the top of each column. Instead of showing all of the blocks every time, only the digit that represents that shape is used as a label.

All of the shapes represent organized arrangements for a bunch of ones; they are just grouped differently. Trading and regrouping (for the 1960s or 1980s crowd) happens so that the same columns can represent the same kind (like terms) of organization in place value categories. By the time the number 7,998 is treated as a total, then all of the columned digits have been converted to ones and then rearranged back into organized columns.

Many hours are dedicated to the concept of adding like terms when your child enters an Algebra I class. The algebra terms such as "x^2," "xy," and "x" come into play. All of these terms include an "x" and look like other terms with an "x," yet they are *not like terms* because the symbols represent different shapes, the same shapes that were used in place value as shown in the table in figure 4.2 and are again explained in chapter 12. Yes, the organization behind place value continues to be just as critical to first grade, second grade, and third grade place value addition and subtraction as it is to algebra.

In algebra when the letters replace numbers, the "x^2," "xy," and "x" in their current "don't know what the number is yet" state cannot be added or multiplied in the usual way; they can only be represented as "a note to add them *after* the numbers get assigned." Letters are used because the number is not "born" (using third-grader Joy's description) or assigned yet, but the x's and y's show the relationship. When the letters *do have numbers assigned to them*, then your child can add, subtract, multiply, or divide them as the relationship "notes" indicate. In the meantime, the "note-to-self" expression with the x's and y's will suffice.

WHAT'S WRONG WITH THIS PICTURE?

One of the first experiences of trying to add or subtract unlike terms comes with fractions. How often has your child added $\frac{1}{4} + \frac{2}{3}$ only to end up with $\frac{3}{7}$ and felt downright puzzled that this answer was marked wrong? Why can't $\frac{3}{7}$ be accurate—they are added, aren't they? Yes, the number digits are added, but the $\frac{3}{7}$ as a *fraction number* is actually less than one of the original fractions being added. Addition is a putting-together action verb so if you put two *positive* number pieces together, then you should end up with a larger piece. The top row of figure 4.3 shows how too many students add fractions without number sense.

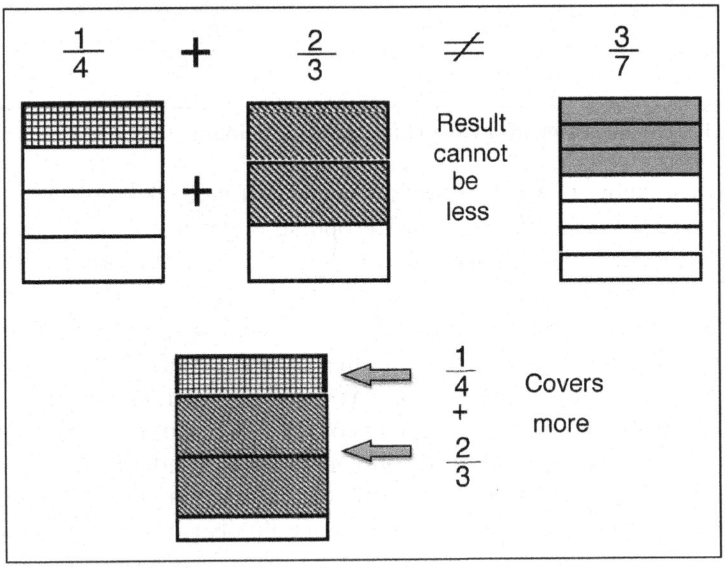

Figure 4.3. What's wrong with this picture?

Let's look at adding fractions with the idea that they can be written in like terms and *share the same category*. Start with the two pieces to add in figure 4.3. In the current expression, the fourths and thirds are not look-alikes nor are they in the same like terms category. Your child can approach this addition with fractions in two ways, and both options will give accurate results. Option 1 uses pictures or models for the fractions and makes them into same size pieces for that like term category. Option 2 uses patterns to find equivalent fractions to show how the two numbers can share a common denominator. Hang on, an explanation is coming!

Neither the one-fourth piece nor the two-thirds piece will exactly fit over the other piece. A commonly shared category must be created so that they can have those necessary "like terms" and share the same category. Place all of the one-fourths pieces on top of all the one-thirds pieces (or trace them and overlap them) like the overlapping pieces shown on the top row of figure 4.4. The second row in figure 4.4 shows how that very small piece in the overlap area turns out to fit over both fourths and thirds.

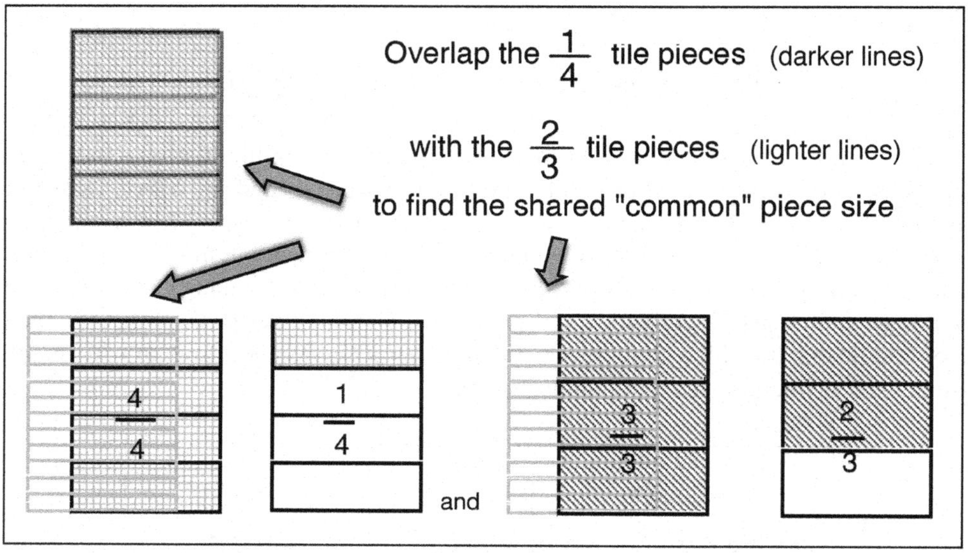

Figure 4.4. Finding like categories visually for adding fractions.

How can you know if twelfths would work if you don't have pictures or models? There are several options to think about. The options might also allow for twenty-fourths to work just as well. The twenty-fourths will have more pieces, but be assured that the twenty-fourths number of pieces *is possible* even though it is not the smallest number category cover. That is what they mean by lowest common denominator (LCD)—the fewest number of pieces in the category for the denominator. By settling on the twelfths as an appropriate common category, your child can *now* add the numerators eight and three to represent the resulting eleven pieces out of the total of the twelfths pieces.

To accomplish the same result using only numbers and patterns, find (or make your own) a pattern equivalence fraction strip for ¼ and for ⅔ and then compare them to find a fraction in both strips that shares the same number in the denominators (same

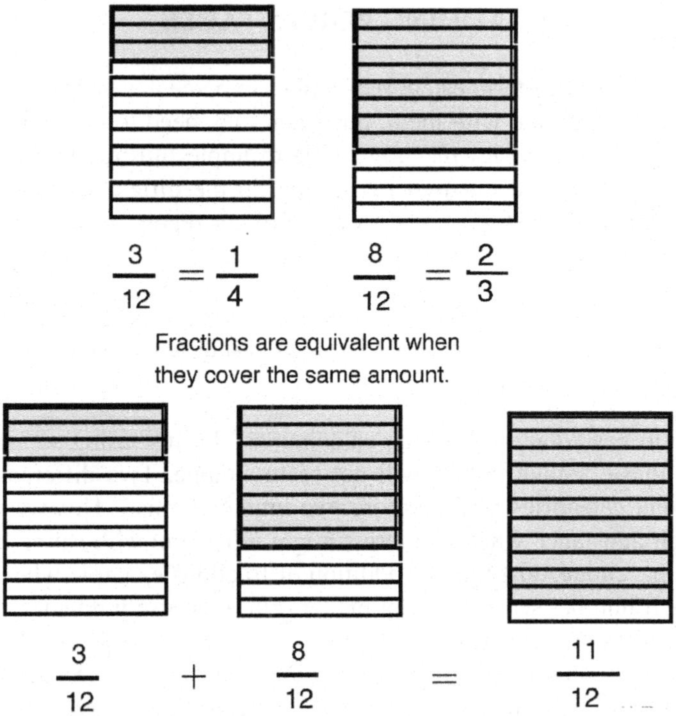

Figure 4.5. Adding fractions after they are in the same category.

category) like the ones shown in figure 4.6. Having the same denominator means that the two fractions are now written in "like terms" and share the same category. With the common category settled, your child can now add the pieces.

When your child knows how to find the new categories of any fractional pieces, then they will feel the same way that Sam felt. Fourth-grader Sam announced that subtraction with fractions should be easy because she now knew how to add them. She must have reasoned that if addition and subtraction were so closely related with whole numbers, then subtraction with fractions must be similarly related to addition of fractions. This kind of thinking makes mathematics so much easier to learn and remember.

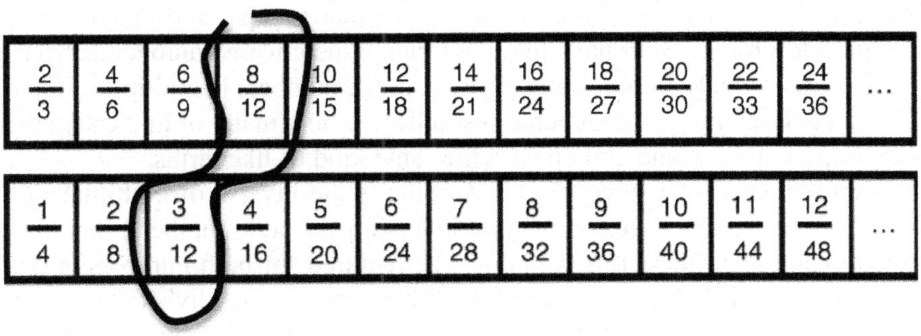

Figure 4.6. Finding like category numerically for adding fractions.

ADDING PERCENTAGES

Another example for like term categories applies to percentages. It is bad enough that your child has to calculate with them, but when they need to combine any of them, they *must keep the categories the same.* One example illustrating how percentages can be misunderstood and misused by mixing up the different categories involves weather forecasting. In this example, a weathercaster reports weather conditions and the chances of rain on a two-day weekend.

> The TV weathercaster announced that there was a 50 percent chance of rain for Saturday and a 50 percent chance for Sunday, and concluded that there was therefore a 100 percent chance of rain that weekend.[2]

If your child has "done percentage calculations" by just using sound-bite "rules" then mixing these "unlike terms" will not seem strange. Two different days results in two different categories and therefore two unlike "terms." This weather example describes a circumstance when like terms might *look right* with sound bite rules, but the result is inaccurate. Just like the addition of fractions example when $\frac{2}{3} + \frac{1}{4} = \frac{3}{7}$, it looks okay if number sense does not matter (but of course it does).

Drawing a picture might help although it is challenging to draw a "Saturday." An analogy using the same relationship idea might work better, like books in a library. Replace "Saturday" with the Salem library (50 percent of the books in Salem library) and "Sunday" with Marblehead library (50 percent of the books in Marblehead library). Does adding those percentages make 100 percent of the books in both libraries? No.

However, combining books within the same library is more reasonable. If 50 percent of the books in the Salem library is calculated and then added to the other calculated 50 percent of the Salem library books (same library so same category), then the resulting amount indeed represents 100 percent of the books in Salem's library. Books in libraries can have more meaning than weather conditions, so now the mistake in mixing the categories is easier to recognize.

A NOTE ABOUT MULTIPLICATION AND DIVISION

Multiplication is about combining numbers in equal groups, and division is about separation into groups. Both have the like terms issue covered automatically if your child thinks like Wendy from chapter 2. If one of the numbers in multiplication identifies the category and the other number just tells you how many of that category, then the resulting number would end up with the same kind of like terms.

Division will work the same way. If division is the inverse of multiplication and having like terms is not required for multiplication for both numbers, then it would seem reasonable that like terms would not be required for both numbers in division either. Wendy would possibly say that one of the numbers in a division problem isn't there the same way either because one of the numbers tells you how to split up the other number, therefore, like terms are already covered.

KEEP IN MIND

Like terms and like categories can make more sense to your child than more rules to remember. The examples in this chapter used several pictures, diagrams, and analogies that are usually preferred by visual learners; however, just enough of the explanation is left out so that the creative child can possibly "fill in the blanks" or design their own methods that grab their imaginations. Join Wendy and Pólya to help your child make the mathematics *really make sense*.

NOTES

1. Pólya, *How to Solve It*, 14.
2. Paulos, *Innumeracy*, 3.

Chapter Five

Artful Assembly of Operations

> All whole number computations are nothing but a sequence of single-digit computations artfully put together. This is the kind of thinking students will need to succeed in algebra and advanced mathematics.[1]
>
> —Hung-Hsi Wu

In other words, place value keeps the digits classified so that they can be "artfully put together." *All numbers* (except Bombelli's imaginary numbers) will fit into the place value organization regardless of how they are generated, calculated, or composed. As soon as a number is written, BAM! (as most students love to shout), it is organized. It is in place value format *just because it is written as a number*. Keeping track of these appropriate columns is about keeping track of the *compositions* of the numbers while *computing* with them, the artful assembly that Wu is talking about. Since numbers come this way, you could say that most of the organization work has already been done.

This chapter describes how different children used their creative imaginations along with artful combinations of single digits to interpret addition, subtraction, multiplication, and division. Either these children did not understand the instructions about the procedures or they were given the opportunity to develop their own ideas and were not, as General Patton cautioned, just told how to do it. Jackie, Zeke, Jake, and Kay are a few of the children in this chapter whose imaginative methods join those of other creative thinkers.

Mathematical procedures, as we know them today, evolved over the centuries and were developed by people who honestly thought that they had the "best way" to calculate a procedure. What is best for one person is not necessarily best for everyone. A particular method may be best for you, or the authors of a book, or even creative children like Jackie or Trevor, but may not necessarily be best for everyone.

FACTS, FROGS, AND FORMATS

The children who described the frogs on the lily pad in chapter 2 understood that when frogs joined other frogs, then the math addition put-together verb should apply. On the other hand, even though George was in the third grade, he was still having difficulty applying his memorized "addition facts" to his current addition problem because he did not know that he had the option to just count in order to get him across the "twenty" mark for the total. He had been told that he could no longer count to solve problems. When Pete applied his "addition facts" to addition problems from chapter 3, the facts only worked sometimes.

Pete and George's efforts to memorize the "addition facts" were successful—possibly too successful. Having spent so much time memorizing "facts," your child, like Pete and George, may have opted to stay with a known fact-based process rather than risk learning a new one. To a risk-averse child in this frame of mind, it makes more child-sense to depend on those facts rather than take the risk of using new, seemingly unrelated information. Children deserve to have new information in math make sense; not just to have the "facts."

Around the same time as those fact "rite-of-passage" drills, a vertical structured format comes into play to begin the formality of an addition procedure. To review the place value column format organization briefly, the single-digit requirement comes from the size of the number represented in each of the place value columns in figure 5.1. Each column is an organizer for a grouped amount, and the digit in each column identifies *how many of that quantity* is in the number. The quantity groups have a pattern where each is one-tenth of the left-hand neighbor as you face the problem. The pattern continues from the smallest nano to those "huge" amounts like googol.

Figure 5.2 is a generalized guide for composing numbers, keeping track of the digits, and providing spaces for meaningful notations. Figure 5.2 shows a general format with different zones for number titles (including the categories from chapter 4), number notations, and number conversions. The problem section has a sample four-digit addition problem that was also used in chapter 4, excluding the decimal titles because they are too small to see. This general format will fit (with a few adjustments to accommodate originality) all of the children's descriptions that follow in the next pages as well as the traditional methods that parents may recognize.

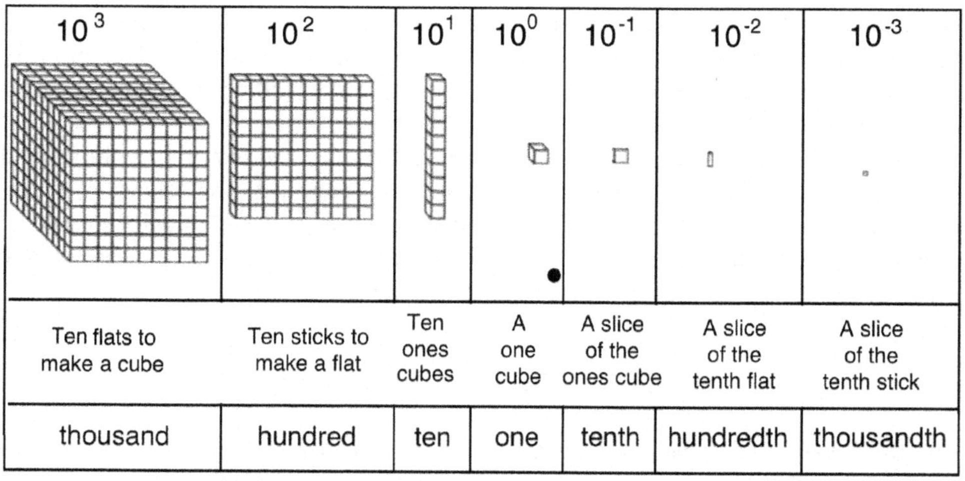

Figure 5.1. Place value blocks and pattern.

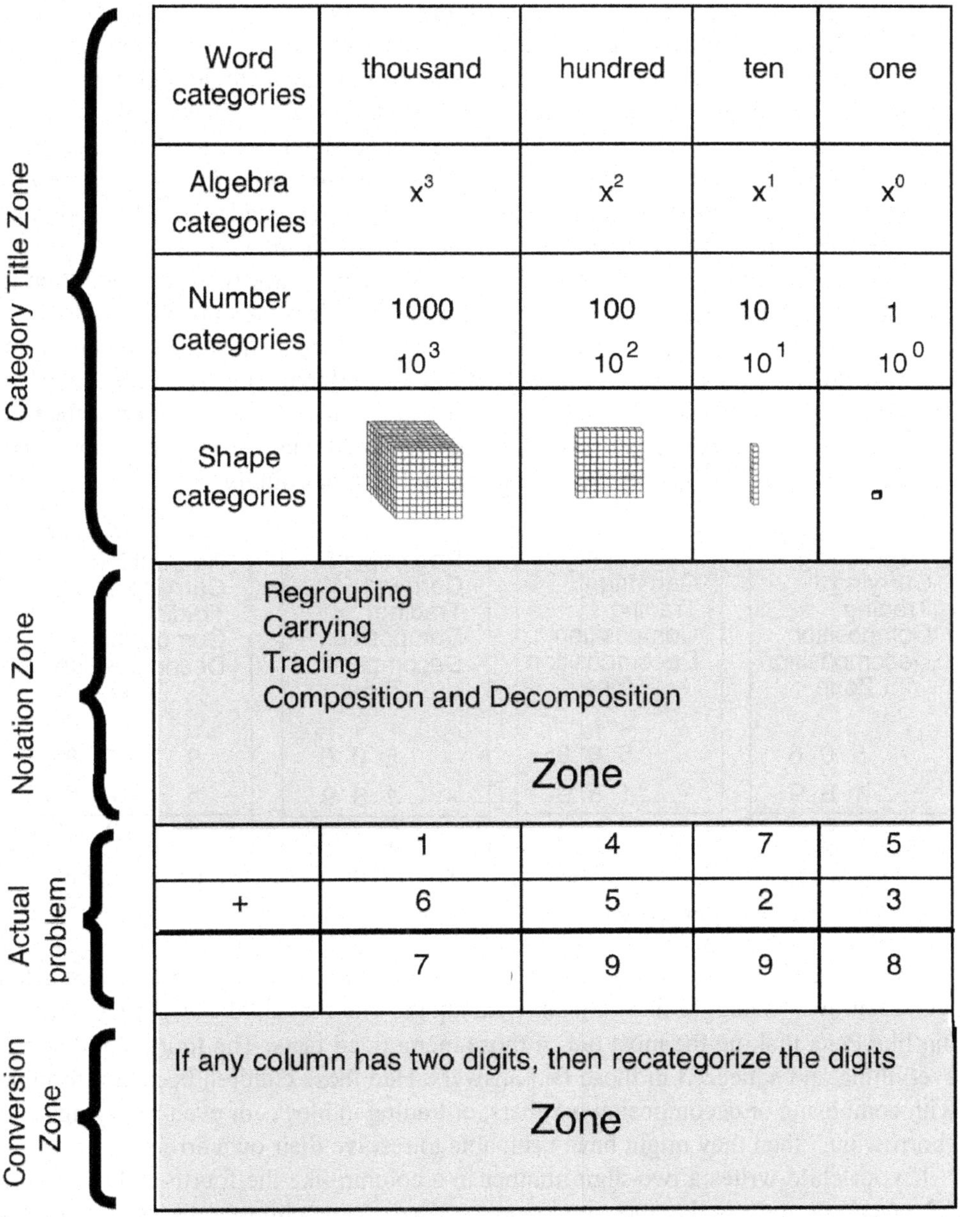

Figure 5.2. Format zones.

Because early calculations were made on small tablets using large numbers written with a cane pen dipped in thin white paint,[2] the limited writing space required that the people doing the calculations should be as conservative as possible with calculation notations. Therefore, instead of using crossing out marks for regrouping notations, the digits were often erased as soon as they had served their purpose. Current paper and computer screens no longer require this limitation, yet we continue to conserve space by using abbreviated notations.

The traditional format for adding, subtracting, and multiplying still makes the most of limited space, but instead of erasing digits, the current notation for regrouping uses slash or cross-out marks followed by rewriting those original single-digit numbers in columns as two-digit numbers. Limiting space and replacing single digits with two digits can prove to be difficult for children to understand because they have been conditioned throughout their short math-lives to *not have two-digit numbers in any place value column.* When this contradiction happens, the notations start to look like squiggles, and are dutifully mimicked, with or without meaning.

If your child is focusing on the "cross-out and rewrite" task with little attention to the actual number sense of what is happening, then their work may look like one or more of the problems shown in figure 5.3. These problems show how the cross-out and rewrite regimen was interpreted by different children.[3] In each case, they are likely just doing what they have been told—or thought they were told! One of the best ways to ferret out the error is to add the "answer" to the number that was subtracted. For example, 318 + 189 is 507 so the error is in the ones column.

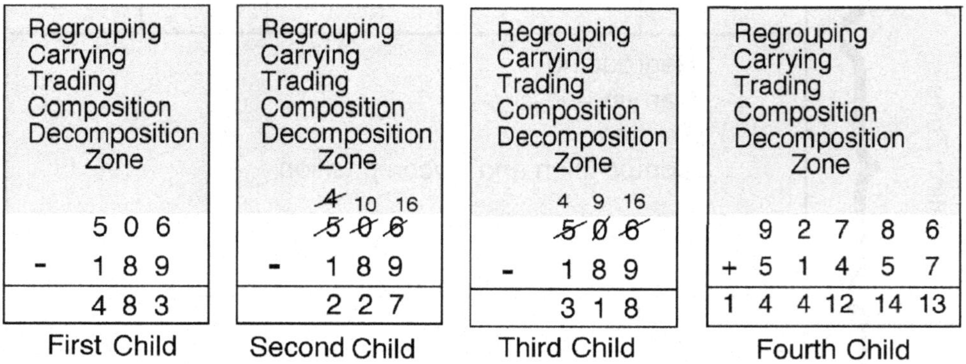

Figure 5.3. Sample of zone errors.

Two of the children dutifully crossed out the numbers on the top row and subtracted with their own "squiggle-like" misunderstandings. The first child was probably thinking like Pete, making the most out of those memorized facts. The fourth child added everything and squeezed in those fact answers. Had these children been comfortable with composing or decomposing numbers, or trading in blocks or even "carrying" and "borrowing," then they might have been able to resolve their own errors.

If your child writes a two-digit number in a column like the fourth child in figure 5.3, please do not say that they are incorrect. These same children, like Pete and others, are confused when you tell them that they are not correct, because, from their perspective, their "fact strategy" worked the first time. Even though they may have been "taught" the regrouping process, the idea does not have much meaning if place value organization is not well grounded. They aren't wrong; they have simply *not finished* with the organization.

These situations provide an opportunity for a parent to have an "organizational guiding moment." The twelve, fourteen, and thirteen two-digit numbers in the fourth child's example just need to be reorganized by size of block (or number section or like terms) so that only one digit occupies any single column. Remember, computation is about artfully combining those *single* digits and maintaining the single digits for the completed like terms organization in place value. Your child just needs, with your guidance, to be reminded of those first-grade experiences when they were categorizing the blocks in the place value columns.

The issue of semantics concerning which terms should be used for changing these digits in columns continues to resurface every few years. As far back as the 1990s and possibly earlier, research results advised against using "borrowing" or "carrying" terminology because of the resulting misunderstanding and ambiguity that children experienced when these words were used in explanations. As Sheila Tobias found in her work about math anxiety, ambiguity is something that *strong minds* resist.[4]

The concept of regrouping numbers (trading, composing, or decomposing) can be difficult enough, especially when the groupings are accompanied by slash marks; why resist a change in math language even if that change might help the language make more sense? Better yet, let children create their own explanations. Most of the children in this chapter's stories did not use slash marks, yet all of them used their own versions of "regrouping." Most children used a left-to-right process like the Hindus,[5] beginning with the larger columns first rather than the traditional right-to-left directional methods that begin with the smallest category columns.

If your child has been composing and decomposing numbers all along, then this trading idea, even with or in spite of the cross-out marks, fits in with what they have been doing previously. To them, the decomposition involved in regrouping and trading is not a risky new idea. Keeping faithful to the assembly of place value organization makes the cross-out and rewrite regimen become what it really is, just somebody's notation of the regrouping exchange.

JACKIE'S THINKING ABOUT NUMBER OPERATIONS

One child proclaimed that what made learning interesting was being able to "think about it."[6] In chapter 1, Jim and Trevor *thought about* their individual creative solutions for multiplication problems. Trevor decided that since the problem included multiplying by "thirty-one," then it was easier to multiply by ten three times *first* before adding those results, and then add the "left over" one from "thirty-one" to the total. Jim had decided to add first even though his decision required that he add a long list of the same number. Both Jim and Trevor understood the single-digit artful assembly requirement.

Jackie, a third grader, devised her own method for subtracting to make sense of the process. Comparing Jackie's method with a "new" procedure called the decomposition method (shown in figure 5.4) helps us recognize common thinking between children's ideas and math procedures. Jackie used the *idea of a negative integer* (possibly thinking like eighteenth-century mathematician Carl Friedrich Gauss who named the "integer") in order to keep the "four" in reserve while finishing her calculations. Thank goodness no one had told Jackie that she could not "take a larger number from a smaller number."[7]

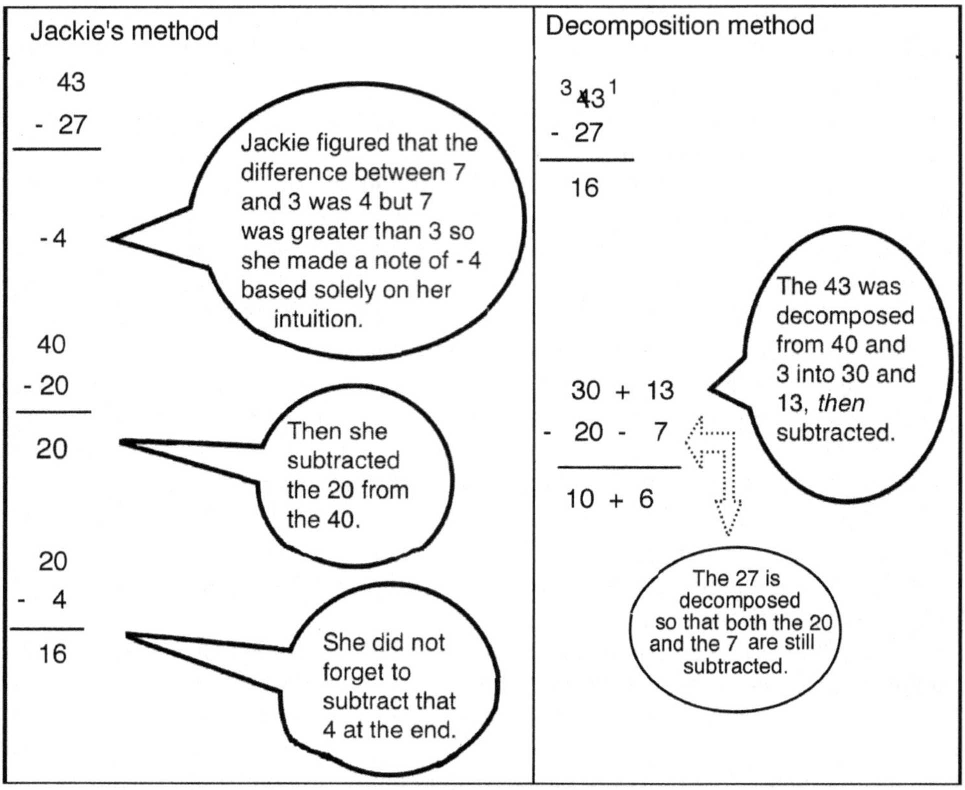

Figure 5.4. Jackie's work compared to the decomposition method.

Three more children's thinking strategies[8] are illustrated in figure 5.5 and are compared to another "new" math procedure called the equal addition method. In all these methods, the children kept faithful to the single-digit-spelled number organized by place value, even though they did not use the traditional right-to-left direction. These three children preferred to calculate using the same direction, much like the early Hindus.[9] Only later, during the twelfth century did another addition method appear in Bhaskara's *Lilavati* that started using the right-hand units first in calculations.

These children, and the Hindus, made sure that the number categories were maintained as they worked from left-to-right, avoiding the dilemma of having two digits end up in one of the place value columns. In other words, the Hindus and these three children, and countless others who intuitively work with the largest part of the numbers first, knew how to avoid confusion and, in the process, make the procedure make sense.

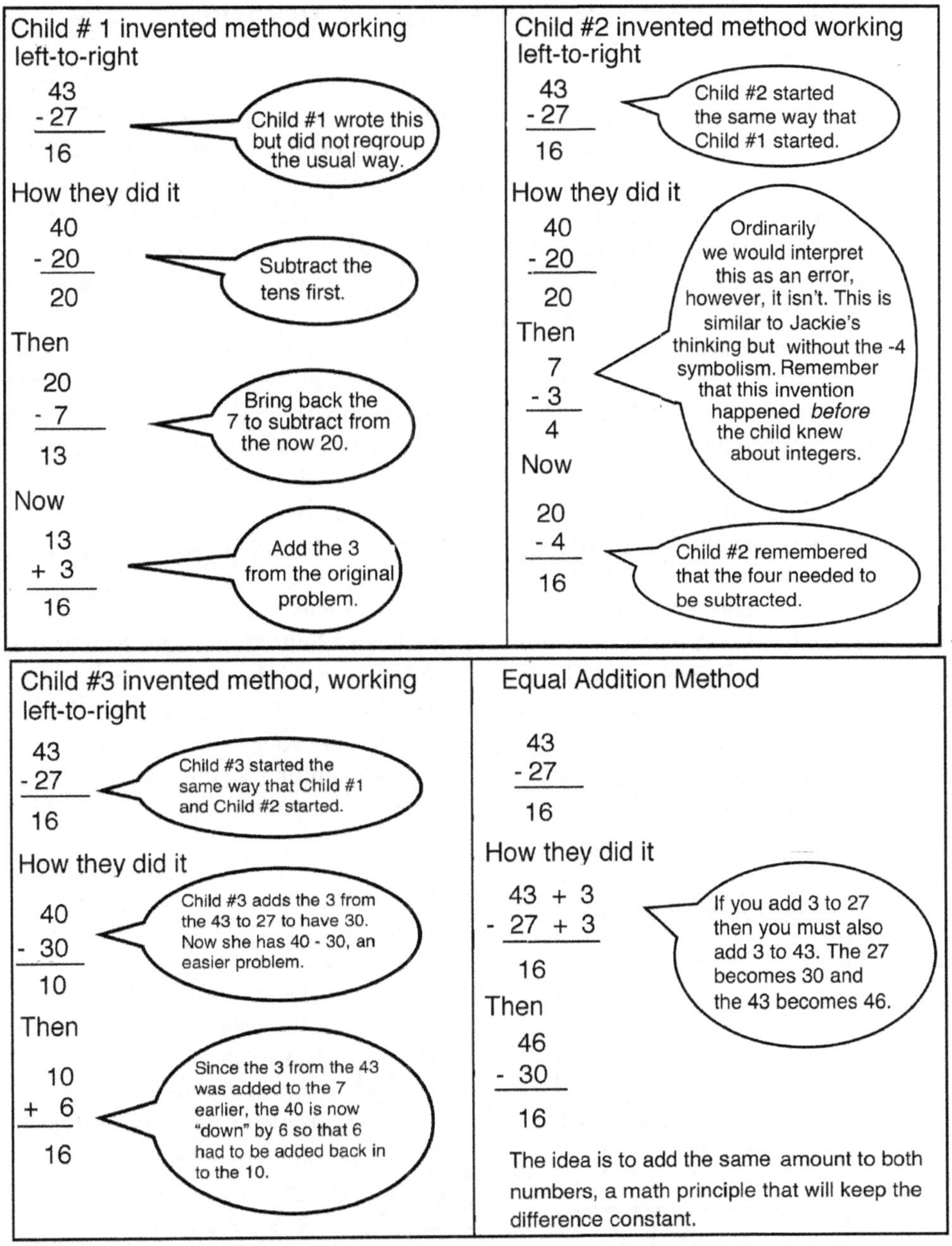

Figure 5.5. Children's explanations compared to the equal addition method.

JAMES, JANIE, AND NAPIER'S LATTICE MULTIPLICATION

James, a fourth grader, had learned the lattice method for multiplication in the third grade and had become very proficient with it because, to him, everything was organized. He understood the single-digits placement in that lattice arrangement illustrated in figure 5.6, probably in the same way that John Napier intended in the sixteenth century. Napier, like James, preferred the inherent organization of the diagonal lines.[10] Diagonal lines separate the sections in the working zones of Napier's lattice arrangement; each diagonal line serves as a place value separator. More about the actual workings of Napier's lattice is explained in chapter 6.

Remember what Wendy said about multiplication in chapter 1? *One of the numbers in the multiplication problem isn't there in the same way that the other one is.* So, according to Wendy, multiplication has one of the numbers "in the air above the other number amount" that must be kept in the imagination because it represents *the number of groups* of the other number. By the same token and concept, when a number is regrouped in place value organization columns, that same idea of a number "in the air" must be kept in the child's imagination in order to maintain the quantities regrouped (or traded or composed) in the columns.

Like James, Wendy had redefined the procedure to make sense to her in her own way, much like Baron Napier,[11] a nonmathematician Math Aficionado. Napier created a grid that had diagonals to keep all single-digit multiplications with two-digit products immediately separated, with the tens digit from the two-digit multiplication placed in a gray zone and the ones digit placed in a white zone.

Janie, a gifted fourth-grade student, was continually putting the digits in her multiple-digit multiplication problems in the wrong location, regardless of how many times

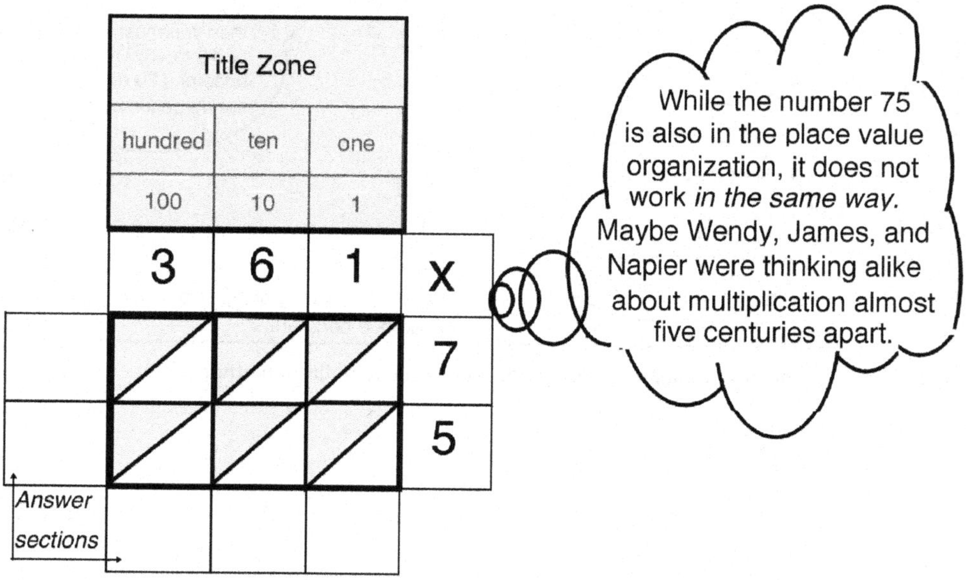

Figure 5.6. Napier's lattice outline.

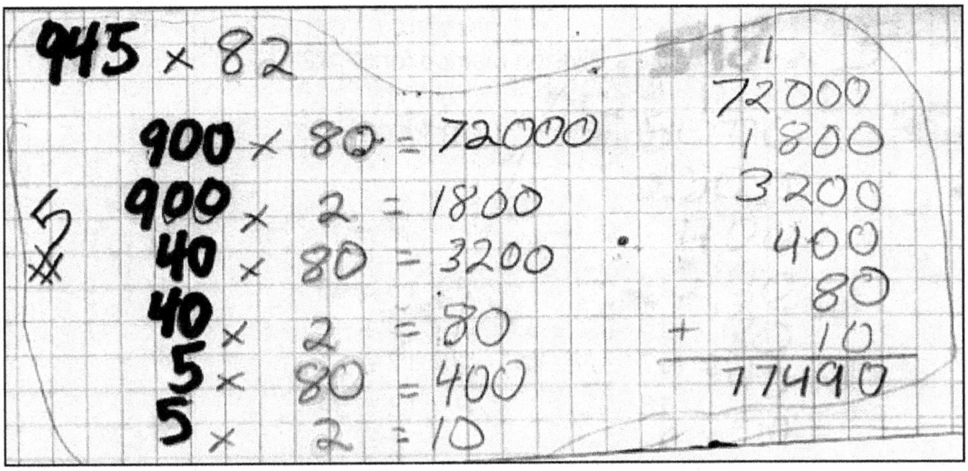

Figure 5.7. Janie's color-coding strategy.

the "correct placement" was explained. She multiplied correctly yet placed the digits in inconsistent locations. *Just telling* her where to place the digits was not working. Then Janie started to color-code the numbers by place value (shown in figure 5.7) with different colors for hundreds, tens, and ones. She used the colors until she was ready to develop her own strategy to keep track of the digit locations. Eventually, she was able to explain her strategy to her teacher's satisfaction.

DIVISION INTERPRETATIONS THAT MADE SENSE

> Ask any fourth grader to explain each step in long division. The answer you are likely to get is, "I don't know why I do it this way. The teacher told us to do it like this."[12]
>
> —Constance Kamii

Why does the very mention of long division make children cringe and bring groans from adults? What makes it so complicated? One issue is reading in several directions at the same time. Which direction comes first? As if this eye-tracking challenge were not confusing enough, you also have to keep track of what goes where—and if none of it bears any resemblance to that multiplication procedure that you just finished, then it is traumatic, indeed.

The guzinta (slang for "goes into") expression is still used in many classrooms and at homework dining room tables, even though "goes into" has no mathematical meaning. Given the division problem, 72 ÷ 3, students learned to ask, "Will three guzinta seven?" If the answer was "no" (even if you knew what "goes into" meant), then you went to the next digit and asked, "Will three go into seventy-two?" With any kind of luck, someone clued you in to what "goes into" meant. That is where the term guzinta comes from; the term still makes no mathematical sense. And, as in Kamii's example, you just do it that way. "Goes into" has nothing to do with real division!

Jake's Division Method for 52 ÷ 3

10	3 x 10= 30
.	.
.	.
.	.
15	3 x 15 = 45
16	3 x 16 = 48
17	3 x 17 = 51
18	3 x 18 = 54

Jake now knew that the number that he wanted was between 17 and 18

17	3 x 17 = 51
⋮	⋮
17.5	3 x 17.5 = 52.5
⋮	⋮
18	3 x 18 = 54

Figure 5.8. Jake's division.

Jake, a fifteen-year-old student, could not get a handle on division, much less long division. When he was asked to explain his own method, his explanation looked like a classic visual-spatial explanation, which is shown in figure 5.8. His method resembled interpolation, a method used for estimating numbers before the dawn of graphing calculators. Just as Jackie had never been "taught" about negative integers, Jake had never been taught interpolation, yet both had created their own versions of a process because neither the regular "trading" (for the 1960s crowd) notations, nor the "guz-inta" method made sense to them.

Jake knew that he needed at least thirty to have three of anything close to fifty-two, so he began with 10 × 3 for his first venture into his estimation process and continued until he passed the fifty-two that was his goal. Jake was strategic about his choices. This is the kind of strategic guessing that George Pólya is talking about! Jake eventually divided that extra one leftover into three pieces so that he could, in turn, multiply with it three times to get the extra one that he needed. Jake's conversion and formalization of his process into a more formal procedure that others could understand is explained in chapter 6.

Kay, a third grader who had a particularly difficult time conforming to traditional place value addition and subtraction regrouping procedures, had found her stride dur-

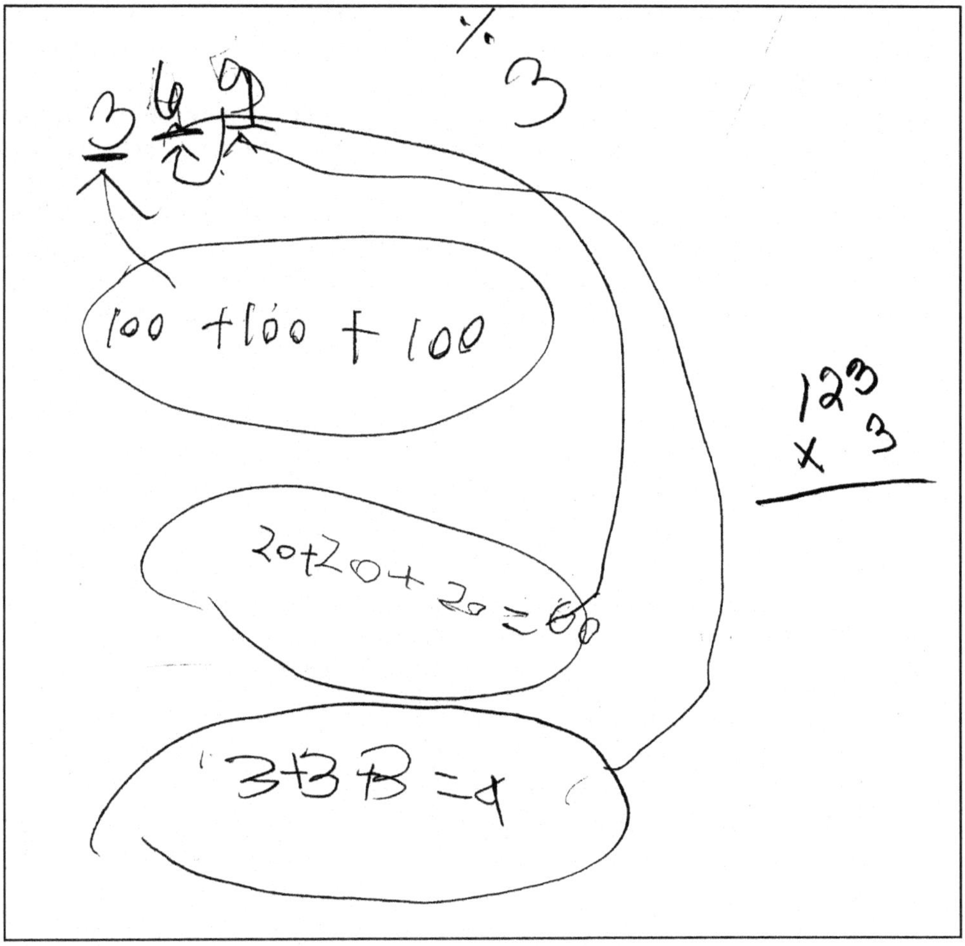

Figure 5.9. Kay's "number cousins."

ing multiplication and division. She had created an addition strategy similar to Jim's, one that added numbers for multiplication. When faced with division, Kay expressed the breaking apart of each place value component as "number cousins." For example, 369 ÷ 3 needed *three* cousins in each of the number columns. She described her cousins for 300 as 100 + 100 + 100, the cousins for sixty as 20 + 20 + 20, and the last nine had 3 + 3 + 3 cousins, all shown in figure 5.9 as Kay drew them.

KEEP IN MIND

The mathematics is in the making sense, the craft, the design, and the thinking. Very often what makes sense to one person does not make sense to another. Of course, it is nice for everyone to understand something the same way, but that does not always happen. Your child does not have to conform to Zeke's reasoning, Jackie's method, Trevor's thinking, Jim's strategy, or Kay's "cousins and partners." What is required, however, is accuracy in the solution and having the math make sense to the person doing the work. Your child and all of the other budding mathematicians are looking for ways to bloom, not just to be told what to do.

Eventually, mathematicians require a formal proof, but for now, an explanation about the reasoning will do, even if that involves "cousins," "partners," negative integers before children are supposed to know how to use them, or an approximation strategy that predates calculators. Be curious about what your child is thinking. Children are eager to learn and very willing to explain their thinking, as long as that "judgment look" is not around. Your child will likely be fascinated that you do not understand something that they do and possibly pleased that they need to explain it to you.

NOTES

1. Hung-Hsi, "What's Sophisticated about Elementary Mathematics?" 5.
2. Eves, *An Introduction to the History of Mathematics*, 183.
3. McKillip, "Teacher Made Tests: Development and Use," 41.
4. Tobias, *Overcoming Math Anxiety*, 50.
5. Eves, *An Introduction to the History of Mathematics*, 182.
6. Boaler, *What's Math Got to Do with It?* 40.
7. Curcio, "There Are No Algorithms for Teaching Algorithms," 25.
8. Kamii, "Children Inventing Their Own Procedures," 202.
9. Eves, *An Introduction to the History of Mathematics*, 183.
10. Boyer, *A History of Mathematics*, 342.
11. Boyer, *A History of Mathematics*, 342.
12. Kamii, *Young Children Reinvent Arithmetic*, 65.

Chapter Six

Same Math, Same Meaning, Different Organization

New vs. Old

> Being able to calculate in multiple ways means that one has transcended the formality of an algorithm and reached the essence of the numerical operations—the underlying mathematical ideas and principles.[1]
>
> —Liping Ma

The scene is a familiar one, a math homework session accompanied by tension, tears, and frustration. Then little Johnny or Jane sympathetically brings mom or dad a handkerchief to dry their eyes. The parents who are often confused and discouraged when yet another way to multiply or, worse, a different way to complete long division appears on a page of practice homework problems. Children plead for sensible explanations and parents are left wondering why "the way they learned it" is not used anymore. One resolution for this situation is to *compare* the different methods, *locate* the digits, and *then ask* "why" both methods work.

These changes get dubbed "new math," and it seems to reenter the conversation whenever another method is presented as part of a different curriculum or math educational initiative. The term "new math" is a reflection of the frustration about change rather than the change in the mathematics teaching. The procedures referred to as "new math" are *not the math*, the procedures are only *methods for creating new numbers*. The *craft* of mathematics is the essence that makes any procedure work. Learning a "new way to divide or add" is not necessary when you can compare and *build a bridge between any two methods*.

Frustrated parents voice their concerns, saying "It took long enough to remember the steps the first time, so why change it?" If anything, understanding different calculating strategies is best, as long as accuracy prevails. It isn't about which *method* is best; the real mathematics is in why different methods *can* make sense. The math procedures being taught in your child's classroom today may look new to you, but, in truth, the actual underlying mathematical ideas and principles behind the math procedures aren't new. Mathematics is in the *design behind the thinking*, and place value keeps the thinking organized.

Procedures may change but place value *never changes*. Place value is the life-blood of all procedures, the reason that they work. Comparing different procedures with place value organization rearrangements is the key to understanding *how* they work and *how* they connect to the procedure that you learned in school. *The locations of the digits may have been changed, but the place value of those digits did not change*. The frustration is often more about the *change in digit locations* rather than the procedure itself. To help keep this frustration at bay, this chapter looks at several procedures, no matter when they were taught, through "place value eyes."

"IF IT AIN'T BROKE, DON'T FIX IT"

Figuring out whether "it's broke" is key. If your child understands the procedure, then it "ain't broke." If your child is asking for help because they do not understand (and neither do you), then it *is* "broke." Rather than "fall in love" with any one particular method, it may work better and be more productive to help your child use their own "place value eyes" to find where the "new locations" of the digits are in the "new" procedure. Forging ahead into regrouping calculations *of any kind* without understanding place value will only lead to more arguments, more frustrations, and more tears.

If your child comes to you for help on one of these new procedures, be very careful to preserve the place value *values* of the digits as you explain or ask them questions about their understanding. Compare the *digit locations* in your preferred method to the new method to find where the place value locations have changed. Both Zeke and Jake made up their own "method" based on what made sense to them and *then* they made their method look as much like the "book" method as much as they could. Whether your child uses the lattice method, partial products, partial quotients, or any other algorithm procedure, place value *must* be preserved.

If it is just about single digits, then why do the procedures seem so difficult to understand? The short answer could be that *they don't make sense to you*. The digits have been moved. When the digits are moved from their familiar-to-you placements, then your equilibrium starts getting unbalanced, or as one student admitted, "it feels like a jelly floor." This isn't a very good feeling, especially when your grade depends on it.

Those *artfully placed* single digits explain how place value comes into play. Keeping the focus on the place value organization in any new method keeps the equilibrium in balance. "There is no algorithm for teaching algorithms."[2] Translation: there is no one perfect way to calculate. Algorithms (methods or procedures) got their start around the tenth or eleventh century, so if you are displeased with a particular method, there's a centuries-long list of people to blame.

All procedures and calculation methods have a road that leads back to like terms and place value organization. The diagram in figure 6.1 shows examples of zoned-outline versions of "traditional" addition, subtraction, multiplication, and division problems. The category titles from the zoned illustrations in chapter 4 have been dropped, except for the title zone in division. The zones are there to focus on where the reclassifications happen. The highlighted fonts for the "traded ones" ("carried and borrowed," "traded," "regrouped," and "composed or decomposed") identify the new locations for the place value classified digits.

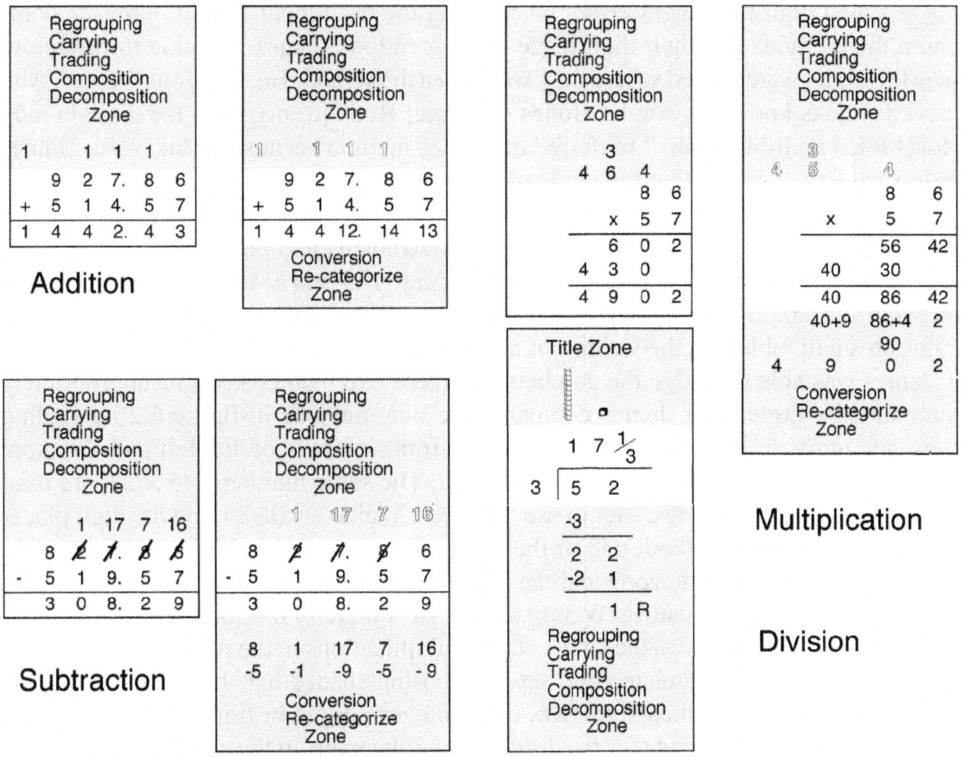

Figure 6.1. All four operations with zones.

Keeping the "carried over" digit (for the 1950s crowd) in the answer space in the conversion zones shows that *the total is still there*, but is just waiting to be rearranged or reclassified. A disagreement continually resurfaces about *where* the regrouped digit should be placed. This kind of argument indicates that the focus is not really on organization but is more on feeling frustrated and unsettled with any flexibility in or changes to digit placement. Keep it simple and sensible. Let your child place numbers *where it makes the most sense to them*. Then let them change it to conform to standard location later—but *with* their understanding.

To paraphrase George Pólya, if you focus only on the arithmetic procedures, whether old or new methods, instead of the relationships within mathematics, then you'll miss the craft of the mathematics. That is the same as focusing on the spelling instead of the storyline in a novel. You will miss the mathematics behind the procedure, just as you will miss the story that the spelled words are trying to convey. You need both spelling and arithmetic to appreciate the full experience of both the mathematics and the storytelling.

MULTIPLICATION ORGANIZATION SENSE

When James entered the fifth grade, he was required to use the traditional multiplication method instead of the lattice method that he learned in the fourth grade, despite his repeated digit placement errors when using the traditional format. After he compared the methods and their digit place value locations, James was able to figure out *why* the digits were placed where they were in each method. He, like John Napier, who carved "sticks known as Napier Bones or Napier Rods [to organize the digits] ready [for] lattice multiplication,"[3] preferred the lattice method because, in his words, James didn't have to "pretend that the numbers changed."

Napier wasn't the first to use the lattice block arrangement. The multiplication lattice arrangement dates to thirteenth-century Arabians and possibly earlier Hindus, and *does not require any crossing out of numbers or erasures* to conserve space. The crossing out notation possibly began with the evolution from writing numbers with paint on small tablets to the writing of calculations on paper.

James was able to bridge the gap between these two methods, yet he appropriately maintained a preference. James compared the two methods in figure 6.2 by finding *the same digits in both methods*. The lattice arrangement is on the left and the more traditional-looking arrangement is on the right. The same numbers, 45 × 23, are used in both formats so that it is easier to use "Where's Waldo" skills to find the digit placements. In the lattice method, one of the multiplied numbers (forty-five) is written on the top with a shaded title zone and the other number (twenty-three) is written on the side without a title because, as Wendy said, it isn't there *in the same way*.

The lattice method keeps the single-digit multiplications in the problem, one row by one column, artfully calculated and separated using shaded half-boxes. *Compare the locations* of the highlighted five, two, one, and zero digits in figure 6.2 to "see" the translation. The regrouping (for the 1990s crowd) happens in the shaded halves. After the single-digit multiplications are completed and placed, they are added on a diagonal outlined in figure 6.3 in order to maintain the place value positions organization. The digits in the resulting "answer section squares" outside of the original block represent place value categories, an artfully put together arrangement indeed!

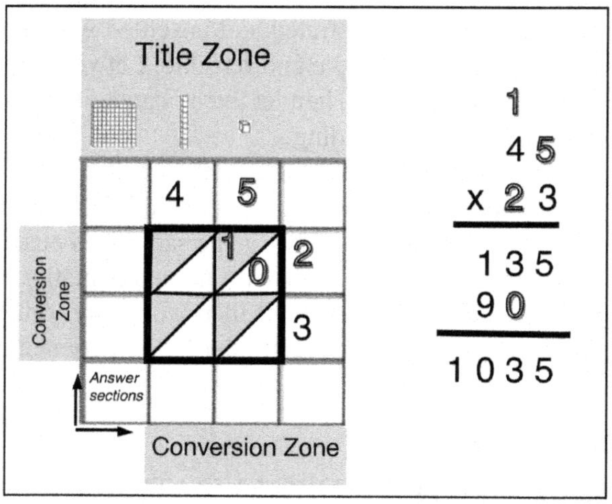

Figure 6.2. Compare the digit locations in the two methods.

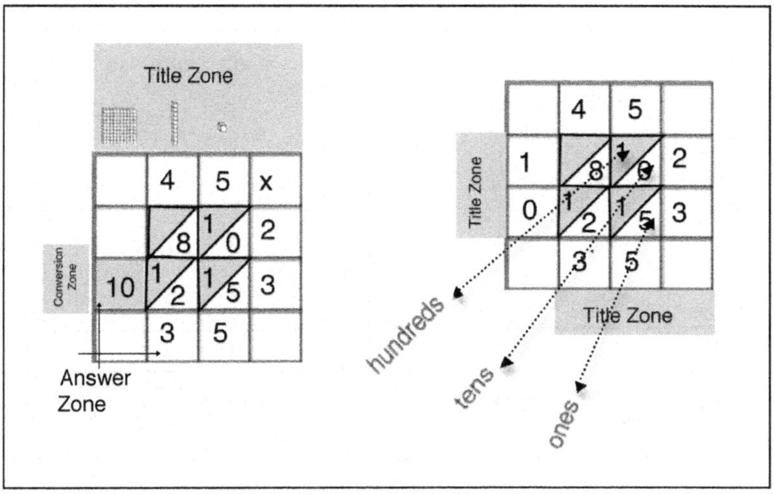

Figure 6.3. Lattice zones showing hundreds, tens, and ones.

The totals in the diagonal "columns" shown with the dotted arrows of the lattice multiplication match with the totals of the vertical columns in a traditional multiplication format. As the numbers in the problems increase, the number of place value columns also increases, and therefore, the number of diagonals increases. The early Hindus, Napier, and James probably preferred this lattice box arrangement with the same passionate preference that some people have for the traditional format for their own reasons, mostly *because it makes sense to them.*

This idea of keeping track of like terms within a multiplication setting is quite sophisticated, especially in the beginning for young learners. When *you* learned the procedure, you may have figured out where to write the numbers after quite a bit of coaxing, and you endured any mistakes until you finally got it right. Your child is jumping through the same hoops, and some children are jumping through these hoops without the benefit of knowing *why* the hoops work the way they do. Keeping the place values sorted can explain the hoops.

DIGIT ALIGNMENT CONTINUES

Did you ever wonder why you had to line up decimal points for addition and subtraction but did not have to for multiplication? An even greater quandary arises when your child must "move the decimal" in a division problem. Decimals do not just move in a number any more than periods move in a sentence. If either of these moves happens, then there is a logical reason. Decimal locations are place value dependent and periods are complete-thought dependent.

Both whole numbers and decimals share the same place value organization, so both kinds of numbers share the same strategies. Addition and subtraction for both kinds of numbers keep faithful to the vertical column alignment. Multiplication and division can *appear* to take on a different organization, but, in actuality, they do

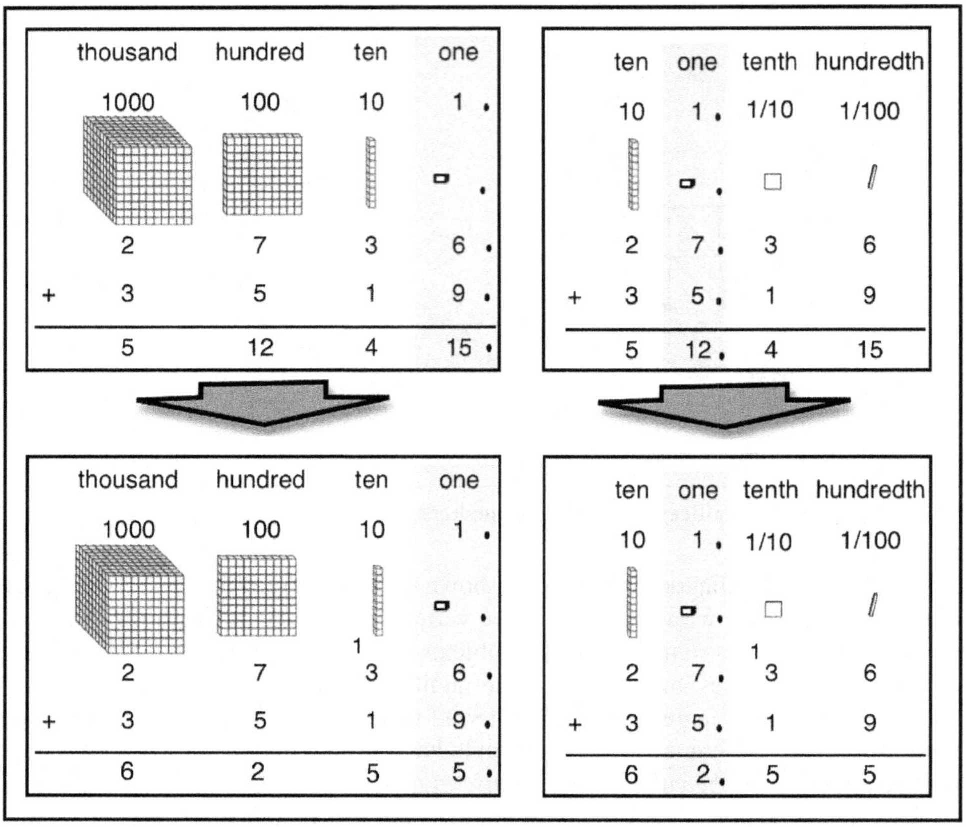

Figure 6.4. Decimal alignment comparison.

maintain the same organization. If your child does not know why the columns are so important to place value organization, then they will continue to place "answer digits" in haphazard locations, probably like Robbie "the faster the better, you get in less trouble that way."

The two examples in figure 6.4 show two typical, traditional addition problems that have the same numbers, but with different decimal locations. Both of these examples require compositions (for the new millennium crowd) and illustrate the artful combinations of the digits in the columns. More digits on the left of the decimal simply mean larger shapes in the pattern and more digits to the right of the decimal will mean smaller shapes in the pattern. The only difference between these two problems is the location of the decimal "period" that marks the end of the whole number sections of the number.

Now compare the traditional multiplication procedure with the partial product procedure in figure 6.5; again, use the same numbers to better compare digit locations when they change. The place value columns stay the same, keeping faithful to the place value *value* and *like term categories*; the conversions and changes happen in their respective conversion zones. The partial product procedure, like the lattice method, does not have crossed out or "canceled" digits, so the traditional procedure *could* be thought of as an abbreviated version of the partial products procedure.

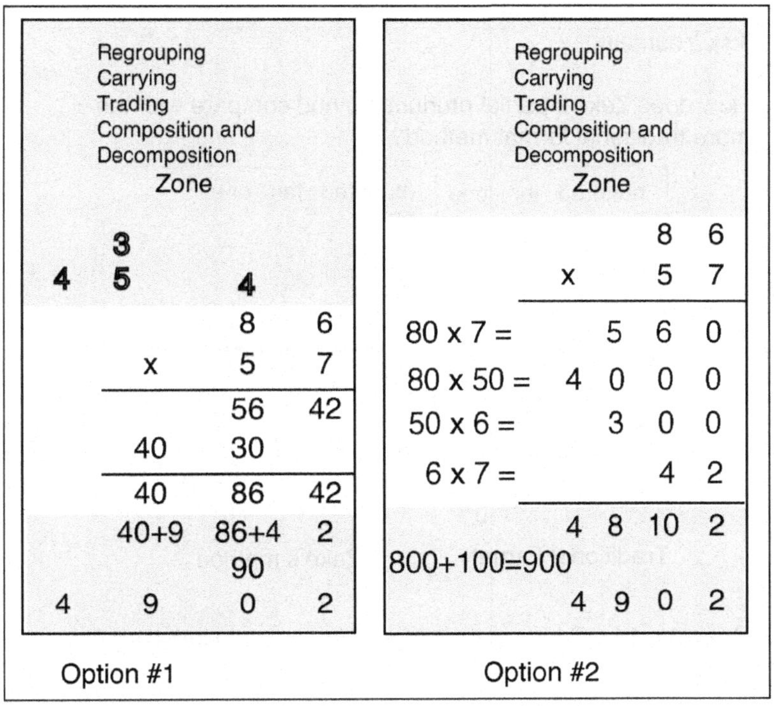

Figure 6.5. Compare traditional and partial product methods.

ZEKE, JAKE, AND LONG DIVISION

In chapter 5, Zeke had learned the valuable "take apart" verb association to division. Now, like James, Zeke was faced with the task of converting his understanding into a procedure that others could understand. The traditional format still wasn't helpful to him, so he used the usual division layout and adapted his process to a more formal procedure that he could remember *and* that others could recognize. It wasn't easy; nor was it quick, but his strategy translation ended up looking very much like the partial quotient format. Two of his stages of thinking are outlined in figure 6.6. The key point is that he preserved place value.

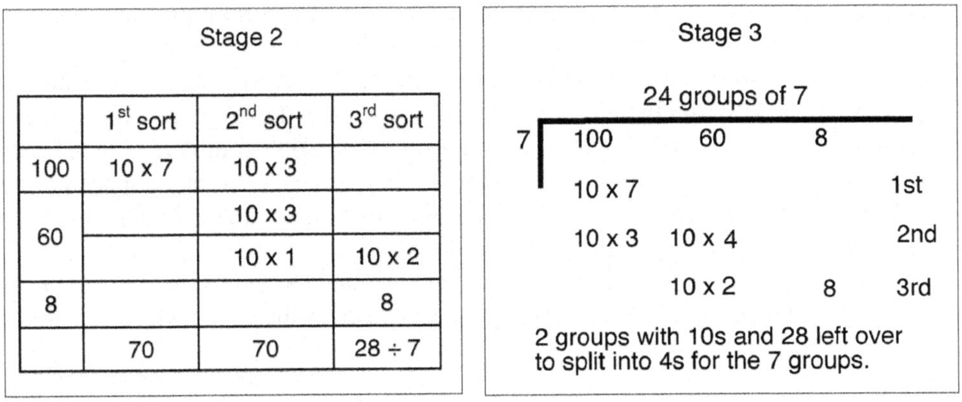

Figure 6.6. Zeke's thinking stages.

Ask Yourself:

How does Zeke's partial product method compare with the more traditional format method?

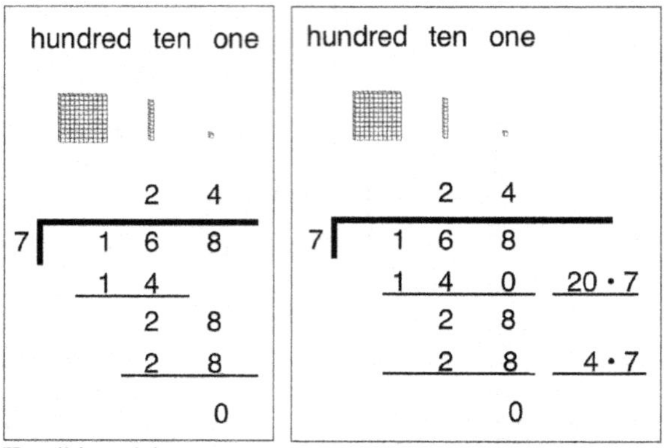

Traditional format Zeke's method

Does the 14 still stay in the same place value location as the 140?

Does the 28 work the same way with both methods?

Which method makes the most sense to you?

Figure 6.7. Ask Yourself.

Jake's invented division-by-estimation process (52 ÷ 3, as described in chapter 5) also needed translation into a more conventional format so that others could understand it. His thinking stages (outlined in figure 6.8) show how he worked through his process. Jake's first efforts to move from his vertical estimation to a horizontal format didn't provide enough exactness for him, so he switched to a strategy similar to Zeke's. Jake's final stage of conversion of his horizontal diagram resembled the partial quotient format. Like Zeke, he understood what was going on *before* he attempted to make it fit another format.

Jake had learned about fractions and knew how to add and multiply fractions, but he had no idea that this situation with division called for using a fraction like the ones he had used in those earlier fraction lessons. He created his fraction notation because it made sense to him. With guidance, and a little retro-connection to place value and exponents, Jake ended up with the diagram shown in figure 6.9, complete with "slices" and fractions, to the amazement and delight of his teachers. Jake had connected the ten slices that made a cube (from place value) with that one-tenth decimal place, so ⅓ of the cube would be about three slices (⅓ of ten slices).

The weeping and wailing befuddlement that surrounds a wide spectrum of issues required by the long division procedure all seem to track back to "why" questions.

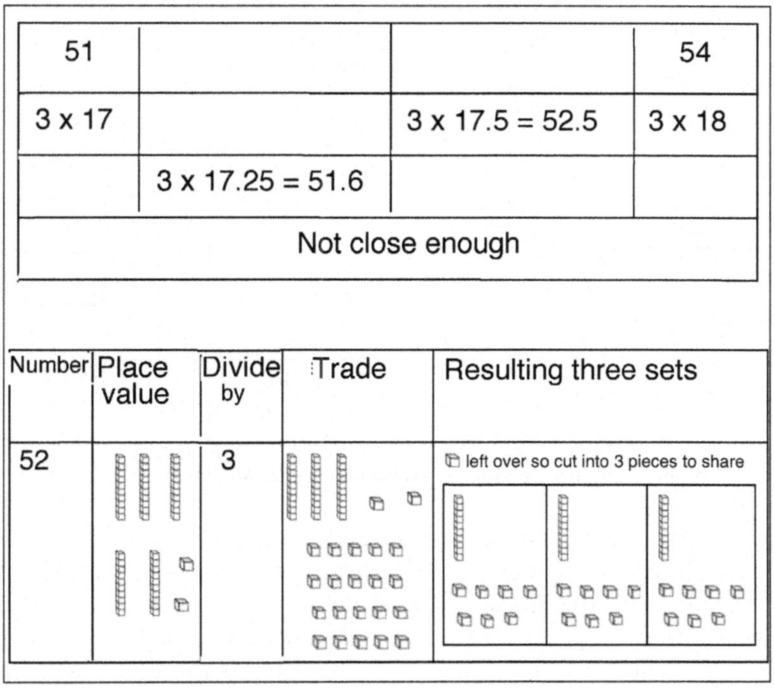

Figure 6.8. Jake's thinking stages.

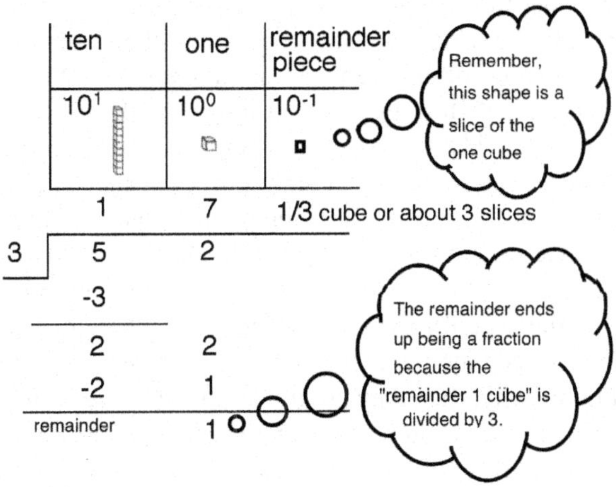

Figure 6.9. Jake's final stage.

Wendy resolved her problem about understanding multiplication, and therefore division, by her declaration that one number served a different purpose than the other. For Wendy, division made it possible for her to figure out the number *in one group* "that she could count," and that the divisor was the elusive "other number." When your child learned that division is the inverse of multiplication, it helped, but that isn't all there is to deciphering the process *or* the procedure.

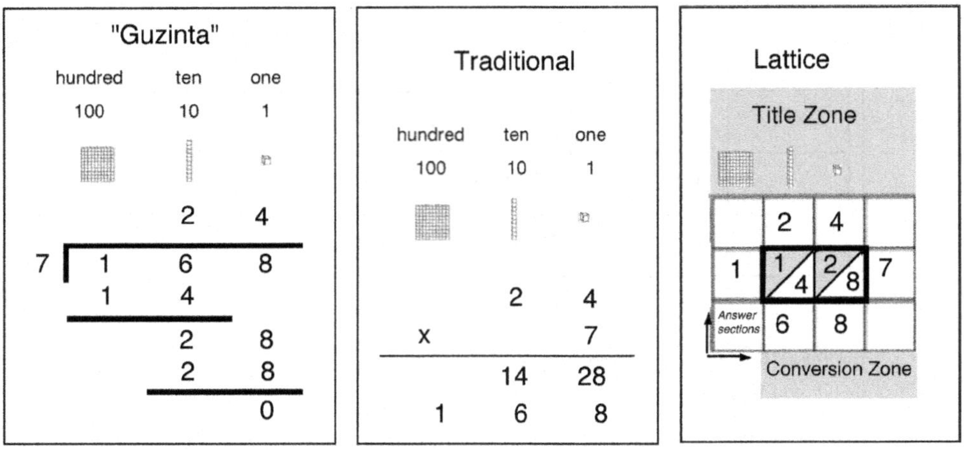

Figure 6.10. Compare the three methods and find where the digits are.

The two multiplication examples and one division example in figure 6.10 show how the lattice multiplication digit or the vertical traditional multiplication digit locations can be compared to the traditional "guzinta" division method. The division procedure is all about keeping the place value organization *organized* and is, believe it or not, just a rearrangement of multiplication. Use your "place value eyes" to compare digit locations with their place value values to try to make some sense of how the multiplication and the division procedures might work together. Compare the different procedures, and conquer the changes in digit placements.

The traditional way of saying that "seven goes into the sixteen" is one of the nonsense nuggets of confusion for children who are already guzinta-challenged, because the one is not large enough for seven to go into. For these children, to magically convert the dividend from 168 to sixteen based on a "one" is nonsense. Remember, the *strong minds* are the ones who reject ambiguity. For the sake of argument and in deference to those parents who grew up using guzinta, try building bridges of understanding by using place value instead of magically changing 168 to sixteen, a numerical error indeed! This is probably why the partial quotient method came into being.

Another problematic issue for the guzinta format is where to put the division "answers." Students may place the numbers where they are told in the name of survival, but if they do not know why, then they are in the same boat as Constance Kamii's student whose teacher just said to do it that way. These questions have answers—when the 168 represents categories of blocks in place value sections it's because one isn't a one anymore; it is one hundred. Six isn't a six anymore; it is a sixty. Zeke just sorted it all out into groups and bypassed the algorithm.

So what is the solution to the long division dilemma? Allow your child to separate first before reorganizing and rewriting into that traditional (or anyone else's methods, including Jake or Zeke) procedure. The separation (sharing, dealing, and other "take apart" verbs) is the key to understanding any of the division format. Just as Zeke was having difficulty remembering where to put the digits in chapter 2, and Jake was befuddled by digit placements, your child might be trying to mimic a procedure and digit placement, too.

Ask Yourself:

Do you see a 14 in both the lattice and the "guzinta" format in Figure 6.10?

Where does it happen? Why do you suppose that would happen?

Do you see a 28 in both the lattice and the "guzinta" format?

Where does it happen? Why do you suppose that would happen?

Figure 6.11. Ask Yourself.

KEEP IN MIND

Your child should choose the method that represents a design or procedure that makes the most sense *to them; or they should make up their own.* What makes sense to Zeke and Jake may not make sense to you, *but it made sense to them.* No one method has the "right of way." As long as the method is consistently accurate, then "it ain't broke." Generally speaking, changes in methods are proposed with the hope and expectation of making the procedures behind the operations make more sense. By comparing different procedure methods using "place value eyes," you and your child might be able to make some personal sense out of some of those newer methods.

NOTES

1. Liping, *Knowing and Teaching Elementary Mathematics*, 112.
2. Curcio, "There Are No Algorithms for Teaching Algorithms," 24.
3. Boyer, *A History of Mathematics*, 342.

Part III

RELATIONSHIPS AND PATTERNS

> The mathematician's patterns, like the painter's or the poet's, must be beautiful; the ideas . . . must fit together in a harmonious way.[1]
>
> —G. H. Hardy

What a concept, harmonious mathematics! Painters have visual patterns, poets have verbal patterns, and Math Aficionados have number patterns. All can fit together in a harmonious way only after they "discover themselves," a phrase that you will read more about in the chapters in this "Relationships and Patterns" section. Recognizing patterns and relationships by thinking through the mathematics that they represent will free your child from the tyranny of memorizing and instead open up the possibility of seeing the harmony and beauty of math.

Students can complete all of the homework without skipping a beat and even pass tests, yet still bypass any understanding of pattern relationships, when they only see the problems as "do it" exercises. Unless they know how to find and identify these patterns, the harmony and beauty of the number relationships stay hidden from their view. Chapter 7 describes a diagonal and vertical viewfinder, a photography term that describes the control of the field of vision. When these viewfinders are placed on math tables, they can help children focus on specific mathematical number patterns by shutting out the surrounding visual clutter.

Chapter 8 expands on the diagonal and vertical viewfinders by including two more viewfinders that can be used with addition, multiplication, *and subtraction and division tables*. The quarter-block viewfinder and triangle-block viewfinder allow for a wider field of vision, extending pattern recognition to rules of number engagements. Chapter 9 is about functions, the predictable equation relationships that reflect the patterns so important to mathematics.

You will read about Joy, a third grader who was trying to make sense of the division symbol and how this symbol seemed to be related to fractions. Her teacher was saying things like, "¾ is read as 3 ÷ 4." Joy saw the line between the two numbers in the fraction and the line in the "÷" symbol, so she decided that the two lines must be related to each other in some way. Her explanation for using the bar in both expressions was that, "in the '÷' symbol, the numbers in the fraction are 'not born yet' because they are represented as dots." The numerator was a "dot" and the denominator was a "dot," so they were not "born yet" in the fraction!

Earlier mathematicians have explored patterns and relationships, like Bombelli with his "wild thoughts" about imaginary numbers or Chuquet's discovery of those "absurd" negative integers, and Abu-Baka's ratios that included Joy's line separating her unborn numbers.[2] Using imagination, looking for patterns and relationships, and making mistakes (and learning from mistakes) makes learning interesting. Children benefit from the new information that results from getting a "no" response, and they will learn that "no" is central to k*no*w, both literally and intellectually.

When children "have permission" to use their imaginations to look at information in different ways and are invited to collect data for their own patterns, they are well on their way to being Math Aficionados. These chapters describe how Kathy, Rudy, and Jesse made their own versions of division, fraction, and decimal tables and how they explained their discoveries *and* their questions. Their questions sparked more tables that caused more questions, making them curious about more tables. Mathematicians formalize their discovered patterns with relationship expressions *after they see* the patterns. With some help along the way, your child can, too.

NOTES

1. Hardy, *A Mathematician's Apology*, 85.
2. Read more about these mathematicians in Eves, *An Introduction to the History of Mathematics*.

Chapter Seven

"Seeing" Math Patterns with Viewfinders

> Perhaps the clearest example of the benefit to be derived from looking at things in different ways is to be found in mathematics.[1]
>
> —Edward de Bono

If you cannot "see" it, then you cannot *see* it. Said another way, if you cannot literally see the math patterns, then you cannot *imagine* more or different mathematical patterns. Not seeing in a literal sense can be caused by "clutter blindness" or figure-ground visual difficulties (a type of visual problem). The cutout windows in the viewfinders in this chapter can help resolve visual difficulties by eliminating the "white noise" distraction of cluttering extraneous information. The viewfinder helps to focus on just the numbers in a given pattern without the extra visual clutter.

Visual figure-ground clutter-blindness happens when the viewer cannot distinguish an item in a picture from the rest of the irrelevant parts of the same picture. They cannot extract the item of focus from the unimportant surrounding visual clutter "noise." Although figure-ground, lack-of-focus difficulties are relatively common in the general population, the extreme version of this condition is very visually disruptive and is described as a special needs learning issue. The viewfinders in this chapter and in chapter 8 can help *all* students focus.

The two viewfinders in figure 7.1 help focus on some patterns within familiar contexts, such as the addition and multiplication tables. Recognizing the pattern is the first task; deciphering and decomposing the pattern into equations is the second. The descriptions in this chapter show only a few examples, but they provide a running start toward helping you help your child understand how number patterns are related to equations and functions across several grade levels and math topics.

The idea is to choose one viewfinder and one of the tables. Place the viewfinder window on top of the table so that only one pattern sequence shows through the viewfinder windows. The square windows in the viewfinders need to match with the squares on the table, so that only one particular number sequence shows in the windows at any one time. The viewfinders are appropriate for all tables, addition and multiplication, as well as a few imaginative ones that Kathy and Rudy created so that they could explore *their own* patterns with fractions.

Diagonal

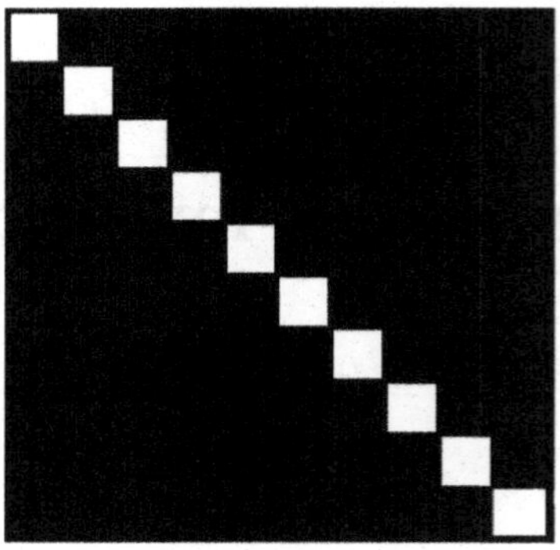

Vertical

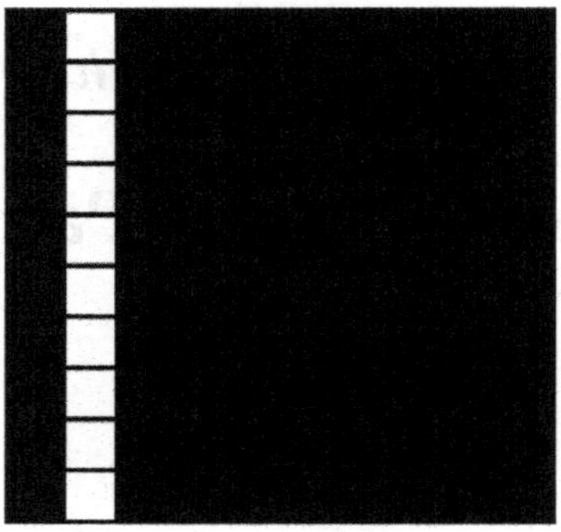

Multiplication Table

x	1	2	3	4	5	6	7	8	9	10
1	1	2	3	4	5	6	7	8	9	10
2	2	4	6	8	10	12	14	16	18	20
3	3	6	9	12	15	18	21	24	27	30
4	4	8	12	16	20	24	28	32	36	40
5	5	10	15	20	25	30	35	40	45	50
6	6	12	18	24	30	36	42	48	54	60
7	7	14	21	28	35	42	49	56	63	70
8	8	16	24	32	40	48	56	64	72	80
9	9	18	27	36	45	54	63	72	81	90
10	10	20	30	40	50	60	70	80	90	100

Addition Table

+	1	2	3	4	5	6	7	8	9	10
1	2	3	4	5	6	7	8	9	10	11
2	3	4	5	6	7	8	9	10	11	12
3	4	5	6	7	8	9	10	11	12	13
4	5	6	7	8	9	10	11	12	13	14
5	6	7	8	9	10	11	12	13	14	15
6	7	8	9	10	11	12	13	14	15	16
7	8	9	10	11	12	13	14	15	16	17
8	9	10	11	12	13	14	15	16	17	18
9	10	11	12	13	14	15	16	17	18	19
10	11	12	13	14	15	16	17	18	19	20

Figure 7.1. Diagonal and vertical viewfinders.

The number patterns showing in the windows are not new; some have been around for centuries, some for only decades, and some patterns are repeated from earlier times, just viewed from a different perspective. The history of these patterns may not matter as much as the benefit derived from learning how to recognize and "see" the patterns in the tables and then deciphering or imagining some other patterns possibly not yet identified.

The multiplication spatial number personality shapes that those fourth graders assigned to the numbers in chapter 3 are again drawn as rectangles in the viewfinder examples in this chapter. The addition spatial number personality shapes might not always be rectangles (unless you want to redraw them as rectangles), because the cutout squares of the number personality shapes are "added on separately," another reminder of Wendy's distinction between multiplication and addition when "the numbers are not counted in the same way."

MULTIPLICATION TABLE PATTERNS

Viewfinders can relieve much of the pain that Math Avoiders can experience with multiplication tables. The pattern of numbers that shows through the windows of the left-

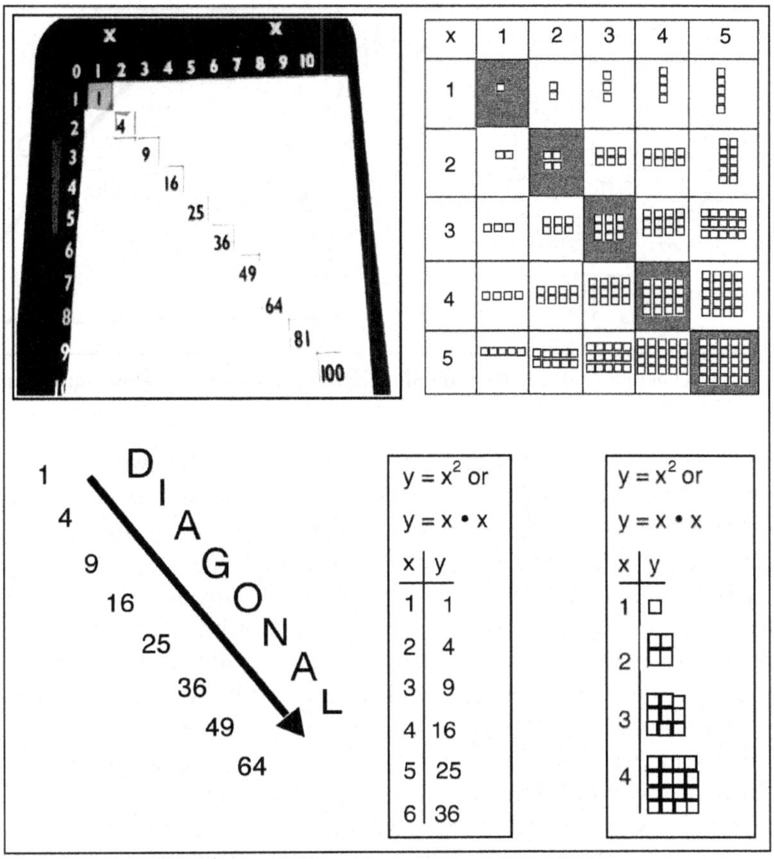

Figure 7.2. Multiplication table with square numbers.

to-right diagonal of a typical multiplication table is represented in five ways in figure 7.2. One of the ways shows a smaller sample of the multiplication table with "spatial number personalities" (from the fourth graders description in the introduction) for the visual-spatial readers. Again, flexibility with expressions will help "different eyes" see "different ways." The reading direction arrows clarify the eye-tracking direction.

The numbers in the viewfinder windows of figure 7.2 are 1, 4, 9, 16, 25, 36, 49, 64, and 81. The *shapes* in the "shape table" represent the square numbers *in the highlighted diagonal* viewfinder windows. The algebraic expression for this square number pattern is x^2 and is expressed as "x squared" (reminders from chapters 4 and 5) because of the square shapes in the diagonal. In fact, you might remember from algebra class that writing square numbers in "squared form" is shown by $1 = 1^2$, $4 = 2^2$, $9 = 3^2$, $16 = 4^2$, $25 = 5^2$, $36 = 6^2$, and so on. More about the algebra for this pattern is explained in chapter 12.

Use the same left-to-right diagonal viewfinder with the same multiplication table and move the diagonal down one row as shown in figure 7.3. This multiplication table pattern may not be as obvious because the shapes may look different. The shapes of the number "personalities" may change, but the number itself stays the same. Remember,

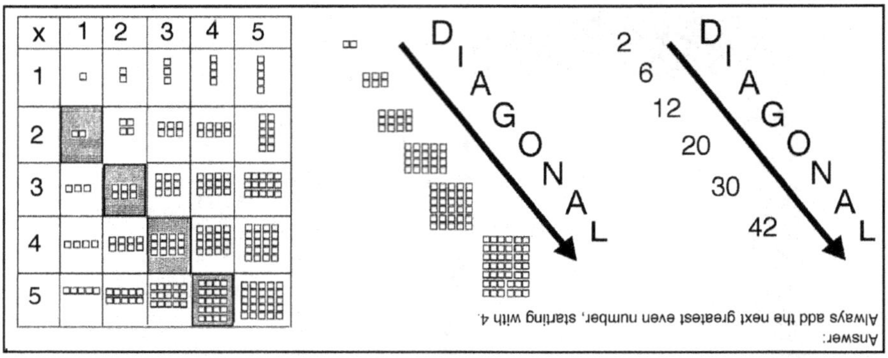

Figure 7.3. Diagonal pattern on multiplication table, one row below square number pattern.

you get to choose and you can take a lead from George Pólya, who suggests, "Mathematics in the making . . . [means that] you have to guess a mathematical [pattern] before you prove it; you have to guess the idea [and] try and try again."[2]

The numbers in the diagonal in figure 7.3 have two more spatial personalities than the grid because the numbers thirty and forty-two are already in a larger multiplication table. The task is to "discover" the pattern appearing in the diagonal numbers. Each new number in the pattern is generated either by adding on numbers or groups of numbers. Imagine what the spatial number personalities would look like for the rest of the multiplication table in this diagonal. Remember, there is flexibility in the spatial number personalities of a number, but the number itself does not change.

Imagine what this multiplication pattern would look like when the multiplication table includes the twelves tables, or even expanded to include multiplications to twenties, thirties, or more. Use the same diagonal viewfinder on different rows, or reverse the direction so that the diagonal windows show the numbers from right-to-left like the diagonals in figure 7.4. Space efficiency becomes a consideration when using the shapes, so after the visual pattern is identified, numbers can take their place. In this example, when the diagonal goes from right-to-left, the pattern stops in the middle and then repeats. The third diagonal shows one possible pattern explanation.

The vertical viewfinder will display one column of windows at a time. Two vertical viewfinder examples are shown in figure 7.6. One viewfinder is on the twos column of the multiplication table and the other is on the fives column. The twos pattern is 2,

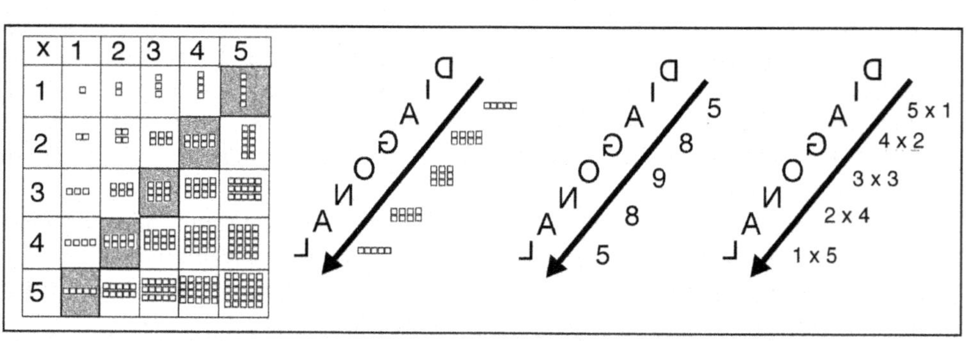

Figure 7.4. Diagonal pattern on multiplication table, in reversed order.

Try This:

Use your best Pólya guesses to find the patterns in the following left-to-right diagonals on either side of the square numbers in the multiplication table.

a. 3, 8, 15, 24, 35, 48, 63, 80

b. 4, 10, 18, 28, 40, 54, 70, ...

c. 5, 12, 21, 32, 45, 60, ...

d. 6, 14, 24, 36, 50, ...

What do you notice is happening?

Do you think this will continue to happen?

Possible Answers:
a. Add odd numbers beginning with 5
b. Add even numbers beginning with 6
c. Add odd numbers beginning with 7
d. Add even numbers beginning with 8

Figure 7.5. Try This.

4, 6, 8, 10 (the even number shapes from chapter 2 return), and on to 20, and the fives pattern shows 5, 10, 15, 20, 25, 30, and on to 50. These same patterns also appear in numerators and denominators of equivalent fraction strips in chapter 4. Recognizing these patterns in other math topics will help your child to *think algebraically*. Some of these patterns are explained "in algebra terms" in chapter 12.

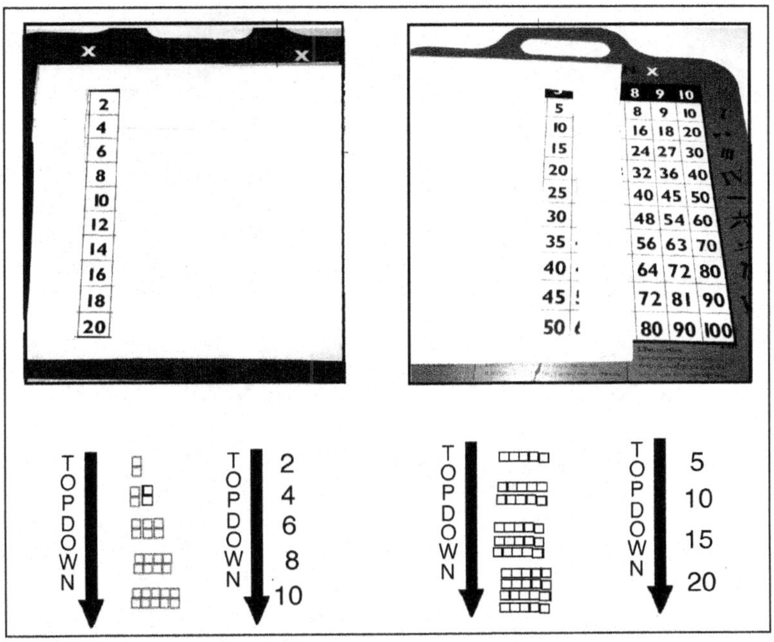

Figure 7.6. Vertical viewfinders for twos and fives.

ADDITION TABLE PATTERNS

The addition table patterns may seem easier than the multiplication table patterns at first; however, both tables have patterns just waiting to be discovered. If number sense plays a part in identifying the patterns, then the addition table and the multiplication table are on more of an equal footing of difficulty. Recognizing how the numbers are composed in these addition and multiplication tables makes understanding the patterns a lot easier, more flexible, and certainly more manageable.

When you place the left-to-right diagonal viewfinder on the center diagonal of the addition table like the one in figure 7.7, the even numbers 2, 4, 6, 8, 10, 12, 14, 16, 18, 20 appear in the windows. The shape personalities for this diagonal pattern appear in the shaded spaces in the diagonal on the right. Those same even numbers appeared in the twos vertical pattern on the multiplication table. Curious Math Aficionados ask "why?" Like Pólya, they try and try again to satisfy their curiosity. When a pattern description makes sense and will accurately apply to other numbers, then you may not need "answers in the back of the book" for confirmation.

Shifting the diagonal viewfinder up or down one level will show odd number patterns. Shift again to another diagonal level and a pattern of even numbers reappears. This odd-even swap also occurs with the multiplication shifts with the left-to-right diagonal. Math Aficionados and mathematicians see these interesting patterns, wonder why they happen, and then start exploring. The exploration and the investigation

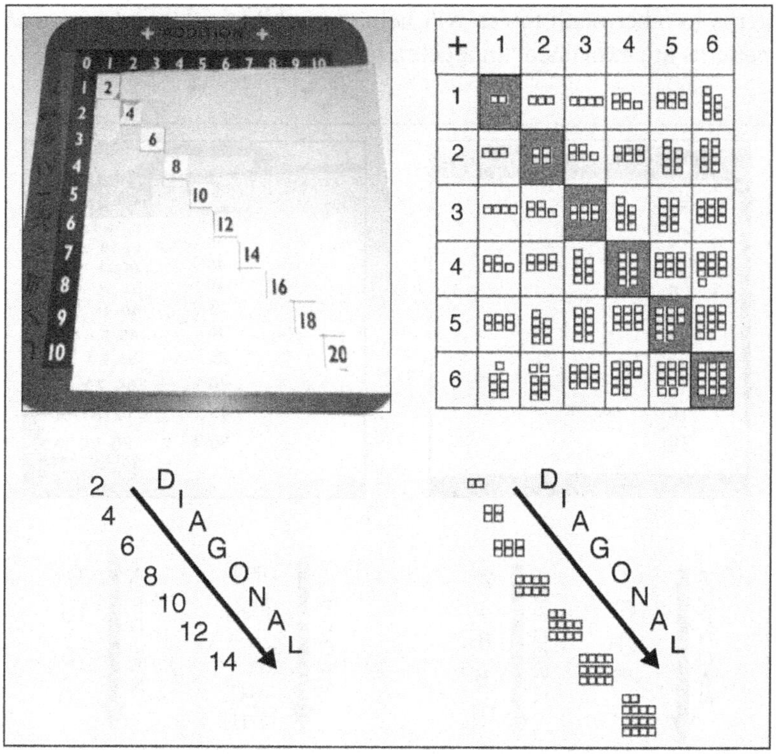

Figure 7.7. Diagonal pattern on addition table with even numbers.

of the patterns are the mathematics that Pólya is talking about that get missed when you stop at the answer.

Paul Halmos says:

> The mathematician at work makes vague guesses, visualizes broad generalizations, and jumps to unwarranted conclusions. He [she] arranges and rearranges ideas and [then] becomes convinced of their truth.[3]

Vertical viewfinders placed on the addition table also show different patterns in the windows with different placements. The same pattern shows in the windows when the vertical column viewfinder is placed in a horizontal position on a row with the same number. Math Aficionados call this repetition the commutative property, and it is special to addition and multiplication because they are the putting-together verb operations. Exploring the division and subtraction tables with viewfinders in chapter 8 will show why the commutative property does *not* generate this same repeating pattern result.

Wrong guesses do not happen when used with a Halmos, or Pólya, intention. Even making a vague guess is still thinking like a mathematician, just don't leave it a vague guess. Pursue and consider and rearrange like Halmos and test for correctness like Pólya until your guess is confirmed or rejected. No matter how your child "sees" the numbers, the pattern message is the same. The mathematics is happening when your

Try This:

Turn the diagonal viewfinder in the other direction to read a right-to-left diagonal on the addition table.

Do you see a pattern?

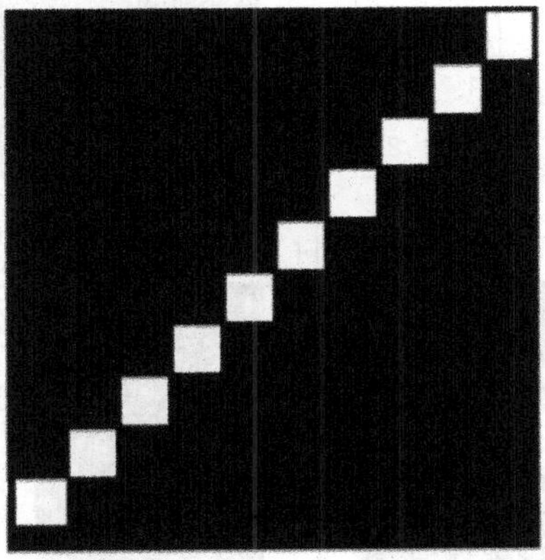

Want to guess a reason why this would be true? Give it your best Halmos guess and then convince yourself that it is true.

Figure 7.8. Try This.

child identifies the patterns; the mathematics is confirmed when your child can express the pattern in words or draw a diagram or illustrate their thinking in a way *that makes sense to them*, and hopefully others.

KATHY'S AND RUDY'S FRACTION VIEWFINDERS

Both Kathy and Rudy separately explored fraction patterns using their viewfinders. Kathy, a sixth grader, made an addition table with fractions, and Rudy made a fraction multiplication table. Kathy was just exploring patterns; however, Rudy was trying to make sense of those multiplication "rules" for fractions that he was learning in seventh grade. Both used calculators to help them generate their respective tables and both decided to use unit fractions for their tables.

Kathy's journal entries describing her viewfinder patterns in figure 7.9 included observations that the diagonal left-to-right denominators decreased and kept the same denominator as the fractions on the edges. The numerator stayed at two for all of the numbers in the window pattern. When she turned the diagonal in the other direction, the denominators increased to the center and then decreased again, a pattern that made her ask questions. Maybe this down-and-up pattern had something to do with those right-to-left patterns in the multiplication tables that also repeated this way.

+	1/8	1/7	1/6	1/5	1/4	1/3	1/2	1
1/8	2/8	15/56	14/48	13/40	12/32	11/24	10/16	9/8
1/7	15/56	2/7	13/42	12/35	11/28	10/21	9/14	8/7
1/6	14/48	13/42	2/6	11/30	10/24	9/18	8/12	7/6
1/5	13/40	12/35	11/30	2/5	9/20	8/15	7/10	6/5
1/4	12/32	11/28	10/24	9/20	2/4	7/12	6/8	5/4
1/3	11/24	10/21	9/18	8/15	7/12	2/3	5/6	4/3
1/2	10/16	9/14	8/12	7/10	6/8	5/6	1	3/2
1	9/8	8/7	7/6	6/5	5/4	4/3	3/2	2

+	1/8	1/7	1/6	1/5	1/4	1/3	1/2	1
1/8	2/8							
1/7		2/7						
1/6			2/6					
1/5				2/5				
1/4					2/4			
1/3						2/3		
1/2							1	
1								2

+	1/8	1/7	1/6	1/5	1/4	1/3	1/2	1
1/8								9/8
1/7							9/14	
1/6						9/18		
1/5					9/20			
1/4				9/20				
1/3			9/18					
1/2		9/14						
1	9/8							

+	1/8	1/7	1/6	1/5	1/4	1/3	1/2	1
1/8				13/40				
1/7				12/35				
1/6				11/30				
1/5				2/5				
1/4				9/20				
1/3				8/15				
1/2				7/10				
1				6/5				

Figure 7.9. Kathy's addition table with fractions.

Kathy started to wonder if these patterns would be different if she used different fractions or even tried making a table for decimals. Kathy's observations generated more questions than answers, allowing her to use many props or fraction models to confirm her guesses, and then ask questions until she convinced herself of the truth. She also used her calculator to help generate a decimal addition table. Kathy was more interested in the patterns *and* in developing her own algorithm. Calculators can calculate very well when the correct keys are pressed in the correct sequence. What *calculators cannot do* is ask questions and *think about and identify the patterns.*

If your child has asked questions about why the $\frac{2}{5}$ is in Kathy's vertical column pattern or if the $\frac{1}{2}$ for $\frac{2}{4}$ is different from the other numerator and denominator pattern in her left-to-right diagonal pattern, then they have not missed the mathematics. The conversation then takes on the task of figuring out *what it should be* if it fit the fraction ratio pattern in the other numerators and denominators. Compare what *they think that the patterns should be* and ask why (and how) the two examples, $\frac{2}{5}$ and $\frac{1}{2}$, could have equivalent ratios. Your conversation can encourage thinking by asking, or reminding, your child about those "like terms" requirements for math.

Because Rudy was trying to make sense of those already strange fraction "rules" around multiplication, his challenge was to find patterns of any kind, the simpler the better. Luckily, Rudy found quite a few repetitions in the numerators of his table in figure 7.10. The fraction denominators were similar to the earlier multiplication

X	$\frac{1}{8}$	$\frac{1}{7}$	$\frac{1}{6}$	$\frac{1}{5}$	$\frac{1}{4}$	$\frac{1}{3}$	$\frac{1}{2}$	1
$\frac{1}{8}$	$\frac{1}{64}$	$\frac{1}{56}$	$\frac{1}{48}$	$\frac{1}{40}$	$\frac{1}{32}$	$\frac{1}{24}$	$\frac{1}{16}$	$\frac{1}{8}$
$\frac{1}{7}$	$\frac{1}{56}$	$\frac{1}{49}$	$\frac{1}{42}$	$\frac{1}{35}$	$\frac{1}{28}$	$\frac{1}{21}$	$\frac{1}{14}$	$\frac{1}{7}$
$\frac{1}{6}$	$\frac{1}{48}$	$\frac{1}{42}$	$\frac{1}{36}$	$\frac{1}{30}$	$\frac{1}{24}$	$\frac{1}{18}$	$\frac{1}{12}$	$\frac{1}{6}$
$\frac{1}{5}$	$\frac{1}{40}$	$\frac{1}{35}$	$\frac{1}{30}$	$\frac{1}{25}$	$\frac{1}{20}$	$\frac{1}{15}$	$\frac{1}{10}$	$\frac{1}{5}$
$\frac{1}{4}$	$\frac{1}{32}$	$\frac{1}{28}$	$\frac{1}{24}$	$\frac{1}{20}$	$\frac{1}{16}$	$\frac{1}{12}$	$\frac{1}{8}$	$\frac{1}{4}$
$\frac{1}{3}$	$\frac{1}{24}$	$\frac{1}{21}$	$\frac{1}{18}$	$\frac{1}{15}$	$\frac{1}{12}$	$\frac{1}{9}$	$\frac{1}{6}$	$\frac{1}{3}$
$\frac{1}{2}$	$\frac{1}{16}$	$\frac{1}{14}$	$\frac{1}{12}$	$\frac{1}{10}$	$\frac{1}{8}$	$\frac{1}{6}$	$\frac{1}{4}$	$\frac{1}{2}$
1	$\frac{1}{8}$	$\frac{1}{7}$	$\frac{1}{6}$	$\frac{1}{5}$	$\frac{1}{4}$	$\frac{1}{3}$	$\frac{1}{2}$	1

X	$\frac{1}{8}$	$\frac{1}{7}$	$\frac{1}{6}$	$\frac{1}{5}$	$\frac{1}{4}$	$\frac{1}{3}$	$\frac{1}{2}$	1
$\frac{1}{8}$	$\frac{1}{64}$							
$\frac{1}{7}$		$\frac{1}{49}$						
$\frac{1}{6}$			$\frac{1}{36}$					
$\frac{1}{5}$				$\frac{1}{25}$				
$\frac{1}{4}$					$\frac{1}{16}$			
$\frac{1}{3}$						$\frac{1}{9}$		
$\frac{1}{2}$							$\frac{1}{4}$	
1								1

X	$\frac{1}{8}$	$\frac{1}{7}$	$\frac{1}{6}$	$\frac{1}{5}$	$\frac{1}{4}$	$\frac{1}{3}$	$\frac{1}{2}$	1
$\frac{1}{8}$						$\frac{1}{24}$		
$\frac{1}{7}$						$\frac{1}{21}$		
$\frac{1}{6}$						$\frac{1}{18}$		
$\frac{1}{5}$						$\frac{1}{15}$		
$\frac{1}{4}$						$\frac{1}{12}$		
$\frac{1}{3}$						$\frac{1}{9}$		
$\frac{1}{2}$						$\frac{1}{6}$		
1						$\frac{1}{3}$		

X	$\frac{1}{8}$	$\frac{1}{7}$	$\frac{1}{6}$	$\frac{1}{5}$	$\frac{1}{4}$	$\frac{1}{3}$	$\frac{1}{2}$	1
$\frac{1}{8}$								$\frac{1}{8}$
$\frac{1}{7}$							$\frac{1}{14}$	
$\frac{1}{6}$						$\frac{1}{18}$		
$\frac{1}{5}$					$\frac{1}{20}$			
$\frac{1}{4}$				$\frac{1}{20}$				
$\frac{1}{3}$			$\frac{1}{18}$					
$\frac{1}{2}$		$\frac{1}{14}$						
1	$\frac{1}{8}$							

Figure 7.10. Rudy's multiplication table with fractions.

tables. He decided to make another fraction table *without using ones in the numerator* so that he could test his Halmos-type guess that all fraction numerators would have "1" when multiplied. His *new table with different fractions* yielded numerators that *were not "1"* but *were* still in his new multiplication table.

Patterns Rudy saw with his viewfinders:

1. The denominators left-to-right diagonals are all square numbers going down.
2. The denominators diagonal right-to-left go up then down again.
3. The square numbers are just like the whole number multiplication diagonal only in reverse order [in the denominator].
4. All of the numerators are one. He changed this when he made his new fraction multiplication table with different fractions.
5. The fraction multiplications in the new fraction multiplication table matched with his area arrangements using fraction tiles (see chapter 10).

KEEP IN MIND

Pattern searching allows children like Rudy and Kathy to make sense of "rules" and procedures, even the fraction procedures. Connecting rules to patterns removes the damage done by the daunting "just do it" command that most students resist, but, regrettably, follow anyway. Your child can use the diagonal viewfinders on diagonals *that do not go through the center* of the tables to find some different patterns or rotate the vertical viewfinders to try to find hints of other patterns.

NOTES

1. de Bono, *New Think*, 125.
2. Pólya, *Mathematics and Plausible Reasoning*, vi.
3. Halmos, "Mathematics as a Creative Art," http://www-history.mcs.st-andrews.ac.uk/Extras/Creative_art.html.

Chapter Eight

Spelunking for Patterns with More Viewfinders

When you start to collect evidence [data] the discoveries make themselves.[1]

—W. W. Sawyer

How nice it would be for the discoveries to do the work. In this chapter, collecting the evidence is about collecting the numbers that represent a pattern. Actually, quite a few patterns are already "collected" in the multiplication and addition tables. These patterns "make themselves" when seen through the viewfinders. Identifying the single-digit number patterns provides a glimpse into the patterns in those assembled multidigit numbers from chapter 6, *and* in Nicholas Chuquet's "absurd" integers, Abu-Baka's ratios, and Raphael Bombelli's "wild thinking" imaginary numbers.

In addition to the diagonal and vertical viewfinders in chapter 7, the quarter-block and triangle-block viewfinders in this chapter help focus on some broader generalities about "rules of engagement" with different kinds of numbers. Identifying patterns is not always easy, but, when the evidence is accurately collected, the discoveries of the patterns make themselves. Flexibility with number expressions makes it easier, or at least more manageable, to express relationships. Collecting information in a spreadsheet or keeping a record in a table gives the patterns a better chance to jump out and identify themselves.

The multiplication and addition tables in figure 8.1 have shaded examples of the triangle-block viewfinder. Both of the sides of the diagonal are readable and are explained in this chapter. A variation of the quarter-block viewfinder on the integer multiplication table exposes only the negative integer section, reminiscent of a mantra in chapter 3 about "two negatives make a positive."

REFLECTIONS ON ADDITION AND MULTIPLICATION

The triangle-block viewfinder placed on an addition table shows two triangular "halves" of the table. Comparing the two halves shows that the numbers repeat. Math Aficionados call that a reflection of the triangle *across the diagonal with even num-*

Triangle-block Viewfinder

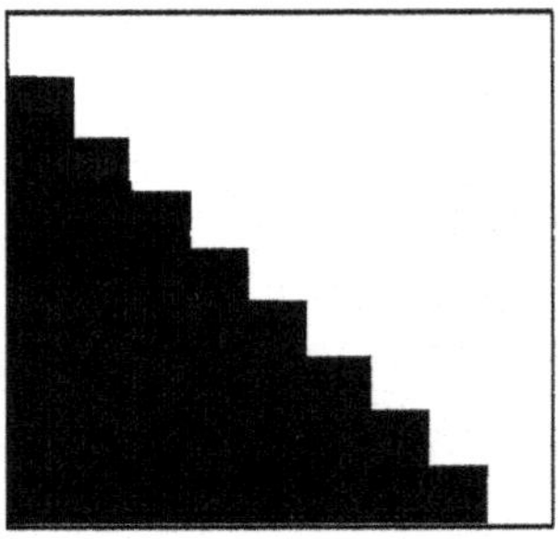

Multiplication Table

X	1	2	3	4	5	6	7	8	9	10
1	1	2	3	4	5	6	7	8	9	10
2	2	4	6	8	10	12	14	16	18	20
3	3	6	9	12	15	18	21	24	27	30
4	4	8	12	16	20	24	28	32	36	40
5	5	10	15	20	25	30	35	40	45	50
6	6	12	18	24	30	36	42	48	54	60
7	7	14	21	28	35	42	49	56	63	70
8	8	16	24	32	40	48	56	64	72	80
9	9	18	27	36	45	54	63	72	81	90
10	10	20	30	40	50	60	70	80	90	100

Addition Table

+	1	2	3	4	5	6	7	8	9	10
1	2	3	4	5	6	7	8	9	10	11
2	3	4	5	6	7	8	9	10	11	12
3	4	5	6	7	8	9	10	11	12	13
4	5	6	7	8	9	10	11	12	13	14
5	6	7	8	9	10	11	12	13	14	15
6	7	8	9	10	11	12	13	14	15	16
7	8	9	10	11	12	13	14	15	16	17
8	9	10	11	12	13	14	15	16	17	18
9	10	11	12	13	14	15	16	17	18	19
10	11	12	13	14	15	16	17	18	19	20

Quarter-block Viewfinder

x		Negative integers						Positive integers					
		-5	-4	-3	-2	-1	0	1	2	3	4	5	
Positive integers	5												5
	4												4
	3												3
	2												2
	1												1
	0												0
Negative integers	-1	5	4	3	2	1							-1
	-2	10	8	6	4	2							-2
	-3	15	12	9	6	3							-3
	-4	20	16	12	8	4							-4
	-5	25	20	15	10	5							-5

Figure 8.1. Triangle-block and quarter-block viewfinders.

bers, the same diagonal of even numbers from chapter 7. Sometimes cutting out the triangle block of numbers (like the example in figure 8.2) works better for kinesthetic hands-on learners. Whether cut out or just compared, the idea is to recognize that the numbers are the same when expressions such as 3 + 4 and 4 + 3 have the same result. This is what Math Aficionados call the commutative property for addition.

The triangle-block on the multiplication table also shows a reflection in the shape, also showing a commutative property for multiplication. This is not a coincidence. Both adding and multiplying are "putting-together" verbs. Look again at the multiplication table in figure 8.1 to see the shaded reflection and the matching numbers. The commutative property applies to single-digit multiplication and continues with multiple digit numbers in an expanded multiplication table. An example excerpt from this expanded multiplication shows fifty-one through fifty-five in figure 8.3, along with some other flexible ways to write fifty-one, fifty-two, fifty-three, fifty-four, and fifty-five.

Be assured that square numbers appear on the diagonal of this table in figure 8.3 even though they may not look square in this form. Those numbers in the shaded row above the fifty-one and fifty-two are called factors because they multiply to generate the fifty-one and fifty-two. The factors cannot be factored into lower numbers so they are called prime numbers; prime numbers are the numbers that can have *just one* spatial number personality rectangle (from that fourth-grade class mentioned in chapter 3) and that rectangle must have one unit measure on one pair of parallel sides.

+	1	2	3	4	5	6	7	8	9	10
1	2	3	4	5	6	7	8	9	10	11
2	3	4	5	6	7	8	9	10	11	12
3	4	5	6	7	8	9	10	11	12	13
4	5	6	7	8	9	10	11	12	13	14
5	6	7	8	9	10	11	12	13	14	15
6	7	8	9	10	11	12	13	14	15	16
7	8	9	10	11	12	13	14	15	16	17
8	9	10	11	12	13	14	15	16	17	18
9	10	11	12	13	14	15	16	17	18	19
10	11	12	13	14	15	16	17	18	19	20

+	1	2	3	4	5	6	7	8	9	10
1	2	3	4	5	6	7	8	9	10	11
2	3	4	5	6	7	8	9	10	11	12
3	4	5	6	7	8	9	10	11	12	13
4	5	6	7	8	9	10	11	12	13	14
5	6	7	8	9	10	11	12	13	14	15
6	7	8	9	10	11	12	13	14	15	16
7	8	9	10	11	12	13	14	15	16	17
8	9	10	11	12	13	14	15	16	17	18
9	10	11	12	13	14	15	16	17	18	19
10	11	12	13	14	15	16	17	18	19	20

3								
4	5							
5	6	7						
6	7	8	9					
7	8	9	10	11				
8	9	10	11	12	13			
9	10	11	12	13	14	15		
10	11	12	13	14	15	16	17	
11	12	13	14	15	16	17	18	19

3	4	5	6	7	8	9	10	11
	5	6	7	8	9	10	11	12
		7	8	9	10	11	12	13
			9	10	11	12	13	14
				11	12	13	14	15
					13	14	15	16
						15	16	17
							17	18
								19

Figure 8.2. Triangle cutouts or triangle-block viewfinder with addition.

	17•3	13•2•2	53•1	3•3•3•2	11•5
X	51	52	53	54	55
51	2601	2652	2703	2754	2805
52	2652	2704	2756	2808	2860
53	2703	2756	2809	2862	2915
54	2754	2808	2862	2916	2970
55	2805	2860	2915	2970	3025
56	[illegible]	2912	[illegible]	3024	3080

Figure 8.3. Example multiplication table for greater numbers and showing factors.

SUBTRACTION, DIVISION, AND MISSY'S DIFFERENT "FAMILY"

Cutting out or just comparing with the triangular-shaped block on the subtraction table in figure 8.4 shows that the commutative "reflection" does not apply to subtraction. However, *there is still a pattern idea.* When the numbers in the triangular-block shaded areas are added together in their respective reflected positions, the result is always "zero," the number on the diagonal! Math Aficionados call this relationship an inverse operation for good reason. That "zero" number earned the moniker of "addi-

a-b		a									
		1	2	3	4	5	6	7	8	9	10
b	1	0	1	2	3	4	5	6	7	8	9
	2	-1	0	1	2	3	4	5	6	7	8
	3	-2	-1	0	1	2	3	4	5	6	7
	4	-3	-2	-1	0	1	2	3	4	5	6
	5	-4	-3	-2	-1	0	1	2	3	4	5
	6	-5	-4	-3	-2	-1	0	1	2	3	4
	7	-6	-5	-4	-3	-2	-1	0	1	2	3
	8	-7	-6	-5	-4	-3	-2	-1	0	1	2
	9	-8	-7	-6	-5	-4	-3	-2	-1	0	1
	10	-9	-8	-7	-6	-5	-4	-3	-2	-1	0

a-b		a									
		1	2	3	4	5	6	7	8	9	10
b	1	0	1	2	3	4	5	6	7	8	9
	2	-1	0	1	2	3	4	5	6	7	8
	3	-2	-1	0	1	2	3	4	5	6	7
	4	-3	-2	-1	0	1	2	3	4	5	6
	5	-4	-3	-2	-1	0	1	2	3	4	5
	6	-5	-4	-3	-2	-1	0	1	2	3	4
	7	-6	-5	-4	-3	-2	-1	0	1	2	3
	8	-7	-6	-5	-4	-3	-2	-1	0	1	2
	9	-8	-7	-6	-5	-4	-3	-2	-1	0	1
	10	-9	-8	-7	-6	-5	-4	-3	-2	-1	0

Figure 8.4. Subtraction table.

tive identity" somewhere along the way, possibly during or following seventh-century Brahmagupta's systematic treatment of algebra, negative numbers, and zero.[2]

Division tables and subtraction tables do not enjoy the same top billing as multiplication and addition tables, even though "division facts" and "subtraction facts" may be part of the memorization ritual. However, both operations can present some interesting "pattern spelunking" investigations. Jesse created his own division and subtraction tables and then started exploring them with those same diagonal, vertical, and block-shaped viewfinders.

Generating division tables and subtraction tables presents an "ordering problem" because the numbers in division do not behave in the same way as they do in multiplication. Both subtraction and division pose the same question, "Which comes first?" As Jesse made his division table, he had to decide which numbers would be the numerator and which would be in the denominator. The numerator number is "being divided into" and the denominator number is "doing the dividing." Just like a driver of a car or the driver software for your computer's printer, the *divisor* is the "driver" for the division. Figure 8.5 shows the vocabulary and along with Joy's third grader symbolism interpretation.

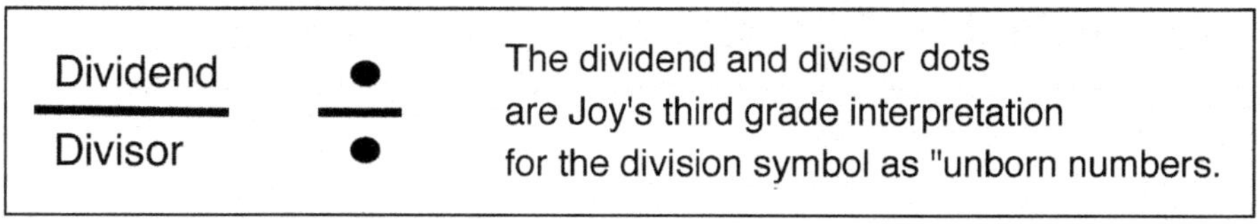

Figure 8.5. Divide symbols and Joy's unborn numbers.

Jesse made two division tables for the numbers one through ten and left them in their ratio form. One table showed the division entries as ratios *before reducing the ratios* and one table showed the ratios *after the ratios were reduced.* Both tables are shown in figure 8.6 to provide the idea of a "larger picture of division and ratios" and to help you and your child find and describe some of the patterns that Jesse found, including some of your own. Use any or all of the viewfinders with these tables to try to find some patterns that "make themselves."

Before reducing

Numerators (dividend number on top); Denominators (divisor number on bottom)

÷	1	2	3	4	5	6	7	8	9	10
1	1/1	2/1	3/1	4/1	5/1	6/1	7/1	8/1	9/1	10/1
2	1/2	2/2	3/2	4/2	5/2	6/2	7/2	8/2	9/2	10/2
3	1/3	2/3	3/3	4/3	5/3	6/3	7/3	8/3	9/3	10/3
4	1/4	2/4	3/4	4/4	5/4	6/4	7/4	8/4	9/4	10/4
5	1/5	2/5	3/5	4/5	5/5	6/5	7/5	8/5	9/5	10/5
6	1/6	2/6	3/6	4/6	5/6	6/6	7/6	8/6	9/6	10/6
7	1/7	2/7	3/7	4/7	5/7	6/7	7/7	8/7	9/7	10/7
8	1/8	2/8	3/8	4/8	5/8	6/8	7/8	8/8	9/8	10/8
9	1/9	2/9	3/9	4/9	5/9	6/9	7/9	8/9	9/9	10/9
10	1/10	2/10	3/10	4/10	5/10	6/10	7/10	8/10	9/10	10/10

After reducing ratios

Numerators (dividend is the number on top); Denominators (divisor is the number on the bottom)

÷	1	2	3	4	5	6	7	8	9	10
1	1	2	3	4	5	6	7	8	9	10
2	1/2	1	3/2	2	5/2	3	7/2	4	9/2	5
3	1/3	2/3	1	4/3	5/3	2	7/3	8/3	9/3	10/3
4	1/4	1/2	3/4	1	5/4	6/4	7/4	2	9/4	5/2
5	1/5	2/5	3/5	4/5	1	6/5	7/5	8/5	9/5	2
6	1/6	1/3	1/2	2/3	5/6	1	7/6	4/3	3/2	5/3
7	1/7	2/7	3/7	4/7	5/7	6/7	1	8/7	9/7	10/7
8	1/8	1/4	3/8	1/2	5/8	3/4	7/8	1	9/8	5/4
9	1/9	2/9	1/3	4/9	5/9	2/3	7/9	8/9	1	10/9
10	1/10	1/5	3/10	2/5	1/2	3/5	7/10	4/5	9/10	1

Figure 8.6. Jesse's division tables.

Numerators (dividend – number on top); Denominators (divisor – number on bottom)

÷	1	2	3	4	5	6	7	8	9	10
1	1/1	2/1	3/1	4/1	5/1	6/1	7/1	8/1	9/1	10/1
2	1/2	2/2	3/2	4/2	5/2	6/2	7/2	8/2	9/2	10/2
3	1/3	2/3	3/3	4/3	5/3	6/3	7/3	8/3	9/3	10/3
4	1/4	2/4	3/4	4/4	5/4	6/4	7/4	8/4	9/4	10/4
5	1/5	2/5	3/5	4/5	5/5	6/5	7/5	8/5	9/5	10/5
6	1/6	2/6	3/6	4/6	5/6	6/6	7/6	8/6	9/6	10/6
7	1/7	2/7	3/7	4/7	5/7	6/7	7/7	8/7	9/7	10/7
8	1/8	2/8	3/8	4/8	5/8	6/8	7/8	8/8	9/8	10/8
9	1/9	2/9	3/9	4/9	5/9	6/9	7/9	8/9	9/9	10/9
10	1/10	2/10	3/10	4/10	5/10	6/10	7/10	8/10	9/10	10/10

Numerators (dividend – number on top); Denominators (divisor – number on bottom)

÷	1	2	3	4	5	6	7	8	9	10
1								8		
2								4		
3								8/3		
4								2		
5								8/5		
6								4/3		
7								8/7		
8								1		
9								8/9		
10								4/5		

Numerators (dividend – number on top); Denominators (divisor – number on bottom)

÷	1	2	3	4	5	6	7	8	9	10
1	1									
2		1								
3			1							
4				1						
5					1					
6						1				
7							1			
8								1		
9									1	
10										1

Numerators (dividend – number on top); Denominators (divisor – number on bottom)

÷	1	2	3	4	5	6	7	8	9	10
1								8/1		
2								8/2		
3								8/3		
4								8/4		
5								8/5		
6								8/6		
7								8/7		
8								8/8		
9								8/9		
10								8/10		

Figure 8.7. Jesse's viewfinder patterns.

The diagonal viewfinder in figure 8.7 of Jessie's pattern searches shows a diagonal of ones instead of the diagonal of zeros that was in the subtraction table. Imagine (or make) a triangle-block viewfinder for Jessie's division table and compare the numbers on both sides of the diagonal. The numbers below and above the diagonal are called "reciprocals" of each other, a reminder of the inverse or "upside-down" description in chapter 2. When these numbers are matched in a reflection-like fashion, paired with their reciprocal, and multiplied, the multiplication will always equal the one on the diagonal, thus the reciprocal has earned the name "multiplicative inverse."

Jesse, like Kathy and Rudy, started asking questions about other division tables with different numbers. He used his calculator to convert all of these ratios into decimals so that he could spelunk for more patterns. Then he reversed the numerator and denominator for another division table and more patterns. Eventually, he used those multidigit numbers for another division table to see if the multidigit table had the same patterns as the single-digit division table patterns. Jesse was experiencing that adrenalin rush caused by success that Andrew Wiles and other mathematicians experience when they shout their versions of Archimedes's "eureka!"

The triangle-block viewfinders on the subtraction table and the division table show reflections, *but the numbers in the triangular blocks are different*, therefore the commutative property does not apply to these two "taking apart" verb operations. A different relationship applies to the "taking apart" tables, giving another perspective on why subtraction and division are described as inverse operations. According to Missy, an eighth grader, subtraction and division are in a different "family" than addition and multiplication.

A FEW RULES OF ENGAGEMENT FOR WORKING WITH NUMBERS

An expanded multiplication spreadsheet that includes both positive and negative integers is provided in figure 8.8. The upper-right section highlights the familiar multiplication table from third grade. After all of the multiplications are completed using both positive and negative integers, more patterns and relationships show themselves. A variation on the quarter-block viewfinder exposes the same positive numbers in both sections, but with different positive and negative integers. Instead of "two negatives make a positive," this rule of engagement can be better expressed as, "multiplying *like signs* with integers provides positive results."[3]

Viewfinders with even and odd number addition and multiplication tables help to focus on some more "rules of engagement." There are some similar "rules of engagement" between even and odd combinations and positive and negative integer combinations. Compare the tables in figure 8.10 to search out your own Halmos guesses about these relationship engagement similarities. Use viewfinders, or make expanded versions of these tables to see more numbers. Use the even and odd personalities and the two-color image "personalities" from chapter 3 to confirm your arrangement "truths."

Generally speaking, the phrases "an even number plus another even number is always even" and "an odd number plus another odd number is even," can be easily recited but panic sets in when someone is asked to explain "why," unless they "see" even numbers as "pairs" as Daniel Tammet described in chapter 3. The same panic

X		Negative integers						Positive integers						
		-5	-4	-3	-2	-1	0	1	2	3	4	5		
Positive integers	5	-25	-20	-15	-10	-5	*0*	5	10	15	20	25	5	Positive integers
	4	-20	-16	-12	-8	-4	*0*	4	8	12	16	20	4	
	3	-15	-12	-9	-6	-3	*0*	3	6	9	12	15	3	
	2	-10	-8	-6	-4	-2	*0*	2	4	6	8	10	2	
	1	-5	-4	-3	-2	-1	*0*	1	2	3	4	5	1	
	0	*0*	*0*	*0*	*0*	*0*	*0*	*0*	*0*	*0*	*0*	*0*	0	
Negative integers	-1	5	4	3	2	1	*0*	-1	-2	-3	-4	-5	-1	Negative integers
	-2	10	8	6	4	2	*0*	-2	-4	-6	-8	-10	-2	
	-3	15	12	9	6	3	*0*	-3	-6	-9	-12	-15	-3	
	-4	20	16	12	8	4	*0*	-4	-8	-12	-16	-20	-4	
	-5	25	20	15	10	5	*0*	-5	-10	-15	-20	-25	-5	

X		Negative integers						Positive integers						
		-5	-4	-3	-2	-1	0	1	2	3	4	5		
Positive integers	5	[redacted]					*0*	5	10	15	20	25	5	Positive integers
	4	[redacted]					*0*	4	8	12	16	20	4	
	3	[redacted]					*0*	3	6	9	12	15	3	
	2	[redacted]					*0*	2	4	6	8	10	2	
	1	[redacted]					*0*	1	2	3	4	5	1	
	0	*0*	*0*	*0*	*0*	*0*	*0*	*0*	*0*	*0*	*0*	*0*	0	
Negative integers	-1	5	4	3	2	1	*0*	[redacted]					-1	Negative integers
	-2	10	8	6	4	2	*0*	[redacted]					-2	
	-3	15	12	9	6	3	*0*	[redacted]					-3	
	-4	20	16	12	8	4	*0*	[redacted]					-4	
	-5	25	20	15	10	5	*0*	[redacted]					-5	

Figure 8.8. Multiplication table with integers and quarter-block viewfinder.

Ask Yourself:

Did you notice all those 0's in the horizontal and vertical rows in the integer multiplication table? Would you like to use your best Pólya guess as to why this could happen?

What are some other "rules of engagement" in the integer multiplication table?

Figure 8.9. Ask Yourself.

Even Numbers

	+	2	4	6	8	10	12	14	16
Even numbers	2	4	6	8	10	12	14	16	18
	4	6	8	10	12	14	16	18	20
	6	8	10	12	14	16	18	20	22
	8	10	12	14	16	18	20	22	24
	10	12	14	16	18	20	22	24	26
	12	14	16	18	20	22	24	26	28
	14	16	18	20	22	24	26	28	30
	16	18	20	22	24	26	28	30	32

Positive Integers

	+	2	4	6	8	10	12	14	16
Positive Integers	2	4	6	8	10	12	14	16	18
	4	6	8	10	12	14	16	18	20
	6	8	10	12	14	16	18	20	22
	8	10	12	14	16	18	20	22	24
	10	12	14	16	18	20	22	24	26
	12	14	16	18	20	22	24	26	28
	14	16	18	20	22	24	26	28	30
	16	18	20	22	24	26	28	30	32

Odd Numbers

	+	1	3	5	7	9	11	13	15
Odd numbers	1	2	4	6	8	10	12	14	16
	3	4	6	8	10	12	14	16	18
	5	6	8	10	12	14	16	18	20
	7	8	10	12	14	16	18	20	22
	9	10	12	14	16	18	20	22	24
	11	12	14	16	18	20	22	24	26
	13	14	16	18	20	22	24	26	28
	15	16	18	20	22	24	26	28	30

Negative Integers

	+	-1	-3	-5	-7	-9	-11	-13	-15
Negative Integers	-1	-2	-4	-6	-8	-10	-12	-14	-16
	-3	-4	-6	-8	-10	-12	-14	-16	-18
	-5	-6	-8	-10	-12	-14	-16	-18	-20
	-7	-8	-10	-12	-14	-16	-18	-20	-22
	-9	-10	-12	-14	-16	-18	-20	-22	-24
	-11	-12	-14	-16	-18	-20	-22	-24	-26
	-13	-14	-16	-18	-20	-22	-24	-26	-28
	-15	-16	-18	-20	-22	-24	-26	-28	-30

Figure 8.10. Addition table for even and odd numbers to compare and finding similarities.

arises when a student is asked about multiplying with even and odd numbers. Use the spatial personality arrangements from chapter 3 to confirm "and convince yourself" of the truth of what you recite.

KEEP IN MIND

Collecting data helps those patterns make themselves. The patterns almost jump out when your child has collected enough information. Your child may ask, "How much is enough?" Enough is reached when your child has the information that they need in order to recognize a predictable pattern. The key to success in mathematics is being able to use whatever strategy works for you and your child to recognize the real mathematics in the patterns. Rules are nice, but patterns are better.

Prove It to Yourself:

Use this multiplication spreadsheet for a selection of odd numbers.

Odd numbers											
1	3	5	7	9	11	13	15	17	19	X	
1	3	5	7	9	11	13	15	17	19	1	Odd numbers
3	9	15	21	27	33	39	45	54	63	3	
5	15	25	35	45	55	65	75	85	95	5	
7	21	35	49	63	77	91	105	119	133	7	
9	27	45	63	77	99	117	135	153	171	9	
11	33	55	77	99	121	143	165	187	209	11	
13	39	65	91	117	143	169	195	221	247	13	
15	45	75	105	135	165	195	225	255	285	15	
17	54	85	119	153	187	221	255	289	323	17	
19	63	95	133	171	209	247	285	323	361	19	
21	69	105	147	189	231	273	315	357	399	21	
23	75	115	161	207	253	299	345	391	437	23	

Do you think you will ever have an even number in this spreadsheet even if you expand it with more numbers?

Create a spreadsheet for multiplying with even numbers Do you think you will ever have an odd number in this spreadsheet if you expand it with more numbers? Use a calculator or Excel if you wish.

Create an addition spreadsheet for integers.

Use any of the viewfinders to help you investigate and explore.

What patterns or rules of engagement can you find?

Figure 8.11. Prove It to Yourself.

NOTES

1. Sawyer, *Mathematician's Delight*, 183.
2. Boyer, *A History of Mathematics*, 242.
3. Sawyer, *Vision in Elementary Mathematics*, 298.

Chapter Nine

Functions, Predictability, and Balance

> The concept of function is of the greatest importance, not only in pure mathematics but also in practical applications.[1]
>
> —Reuben Hersh

Predictability is the life-blood of mathematical relationships, and is important enough to have its own category as an equation called "function." Lack of predictability leaves results to chance, the haphazard kind of chance that isn't very useful. Being able to predict results leads to a kind of stability, a dependable relationship that has a functional purpose. Not every relationship has an equal balance, nor is every equation a function, so it is helpful to learn how to tell the difference. This chapter is about the importance of predictability and the usefulness of flexibility, and how they both help to debug mathematical misunderstandings and sidetrack confusions.

The idea of balance is paramount for equations; having inequalities is also important as long as the necessary conditions are met. Physical balancing tools help visual-mechanical learners to get a *feel* for how the equal sign (=) affects equations. "Any equation whatsoever is no more than the putting down of two different ways of describing something. Yet the usefulness of having two ways instead of one is so great that it is one of the cornerstones of mathematics."[2] In mathematics, those predictable and dependable relationships between two quantity expressions are, not too surprisingly, called functions.

PREDICTABILITY AND FUNCTIONS

Functions and formulas are actually pattern relationship descriptions in disguise, just written in equations instead of words. The predictability in these functions means that one number cannot generate two different results. The equal sign appears in functional equations because of that necessary predictability feature. As Liping Ma's teacher noted, "The equal sign is the soul of mathematical operations. In fact, changing one or both sides of an equal sign for certain purposes, while preserving the 'equals' relationship, is the 'secret' of mathematical operations."[3]

ONE ANSWER OR MANY ANSWERS, SAME BALANCE

Keeping a record of which numbers balanced led a first grader named Hilary to discover how several weights on numbers made a difference in the balance of those numbers on a number balance. She was also intuitively learning about the physics of balancing weights when the weights were placed farther away from the center. Hilary's dad made a homemade number balance like the one shown in figure 9.1, and used equal weights to help her get a feel for balancing "the weighted numbers" as she played with balancing numbers with addition (combinations of single weights) and multiplication (several weights on one number) on the balance.

One of the number questions presented to Hilary was, "how can you use five weights on the left to make a balance for two weights on the right?" Hilary tried several different positions before finding a balance with five weights in the "one hole" on the left side and one weight in the "three hole" and one weight in the "two hole" on the right. She then tried again with four weights in the "one hole" and one weight in the "four hole" on the left side that balanced with two weights in the "four hole" on the right side. Longer arms on a number balance allow for more number combinations and balances. Algebraic versions for some of Hilary's explorations appear in chapter 12.

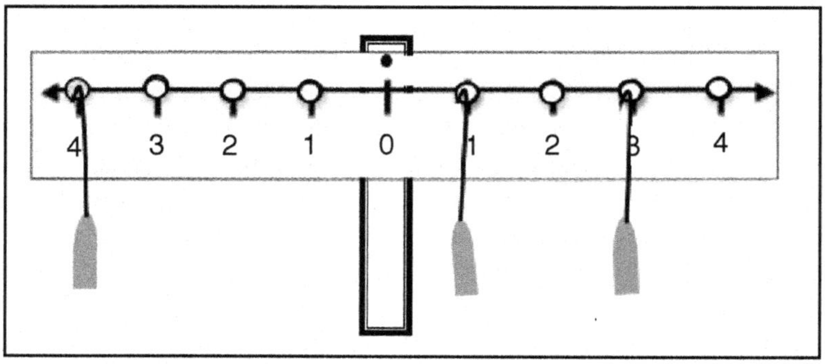

Figure 9.1. Hilary's dad's homemade balance.

This kind of number balance did not help Jeannie, a ten-year-old who was still having difficulty with balancing simple equations. To Jeannie, the numbers did not represent anything tangible for her to balance, possibly because the numbers were still too abstract. An equal pan balance scale (like the one shown in figure 9.2) used in science experiments can balance objects that can translate to equations and number balances. That "=" sign between equation expressions is like a fulcrum balancing two forces. Just "moving anything to the other side" in an equation will upset the balance, and the teacher.

Jeannie's education evaluation report recommended that any math instruction should include experiential and hands-on learning because she had been identified as having "very strong visual-perception skills." Jeannie used the equal-pan balance scale to make some balanced block combinations and kept a record of the blocks that balanced. Next, she drew pictures to represent the balances, and lastly she drew a balance diagram with numbers. Eventually Jeannie was able to write equations similar to

Written example for the equal pan balance:

$5 + \square = 2\square$

Translation was later written as:

$5 + x = 2x$

Figure 9.2. Equal pan balance for Jeannie.

the examples shown in figure 9.3. Your child might be like Jeannie, a picture does not make sense without the equal-pan balance experiences.

Solving equations begins early in math learning. The equations start simple, like the equation on the left in figure 9.3, and then increase in complexity by including more than one option, like the equation on the right. However, remember that all of the equations, simple to complex, require keeping a balance between both sides of the equality (=) symbol. The more complex equations allow several pairs of numbers that maintain the balance, even when twelfth-century Abu Bakr al-Hassar's fractions and fifteenth-century Nicolas Chuquet's absurd negative integers are allowed.

A simple equation with one option must balance:

$3 + \_\_\_ = 5$

$3 + \square = 5$

eventually becomes

$3 + x = 5$

Same number keeps all three equations balanced. Algebra solution is x = 2.

Same equation with two options must still balance:

$3 + \square = \triangle$

eventually becomes

$3 + x = y$

Several pairs of numbers can keep this equation balanced. If x = 2, then y *must* be 5, or if x = 4, then y must be 7. The numbers change but the balance stays.

Figure 9.3. Two equations: one answer or many answers, both must stay balanced.

The simple equation on the left in figure 9.3 is just one of the options for the equation on the right to stay balanced, like going from a specific case with one number to a more general case with several numbers. When these equations are translated into words—those word problems that Math Avoider's dread—they give story backgrounds to the need for balance. Stories can be simple or have lots of detail. When children make up their own stories, their imaginations and fantasies make the stories more alive to them.

A word description for the equation example on the left in figure 9.3 could be, "I have three apples and I need to have five for my friends. How many more apples do I need?" A more complex problem with more words for the equation on the right could be, "I have three apples and I want to know how putting more apples in the bucket will change the total number of apples in my bucket?" Admittedly, not many people will ask this kind of question with apples, but many people involved with managing inventory for very large warehouse "buckets" are faced with a more complex version of this metaphorical apple-gathering example.

A more pragmatic complex problem with even more words and more conditions and different constraints could be, "I want to have a party and give an apple to everyone at the party. Each apple costs $0.45. How many friends can I invite if the decorations already cost $10.00 and I only have $50.00 to spend on the whole party?" There are several pragmatic ways to solve this apple problem. One of the ways is with an equation like the ones in figure 9.4. Another way is to first mentally subtract the $10 from $50 and then approximate the number of apples that $40 would buy. Using $0.50 instead of $0.45 allows a quick estimate of eighty apples for $40.

Taking the equations route would require an accompanying record to list the number of apples and the related costs, like example in figure 9.4. Zeke would likely have done the calculations mentally while others may have wanted more structure. The "best" solution depends on the need for exactness and the details of the question. The words, the conditions, the constraints, and the story context can change, but the balance and predictability must remain in the equation in order to be a function.

Several equations will fit this situation:

$y = \$0.45\,x + \$10.00$; $y - \$10.00 = \$0.45\,x$

$y - \$0.45\,x = \$10.00$

Need to figure out how many friends you can ask and still stay under $50.00. Keep a record to keep track of the costs.

You can invite 88 friends and stay within budget.

If you only depend on calculator without any of the thinking, then you would invite 88.88 friends.

Wonder what 0.88 of a friend would look like?

x	y
1	$10.45
2	$10.90
3	$11.35
⋮	one at a time will take too long
10	$14.50 still too small
⋮	
100	$55.00 too much
90	$50.50 still too much but not by much
88	$49.50

Figure 9.4. Several equations match with the same table.

The solution does not appear in the record until the $50 limit is reached. The eighty-eight friends becomes the one piece of information that conforms to the $50 limit since eighty-nine invited friends will go over budget. Several forms of the same balance relationship are also just as balanced as the equation in figure 9.4. If a change is required, the real mathematics happens when the appropriate inverse operations are used to keep the equation balanced while solving the equation. More information about the algebraic expression of this example appears in chapter 12.

Neil, a gifted second grader, used some of the equation mats like those in figure 9.5 to explore combinations for numbers with a deck of digit cards. He confirmed his equations on the mats by using blocks on the side. He had either already used a number balance or an equal-pan balance to understand balance, or he was one of the children who had acquired number sense just by using blocks. The deck had multiple cards with the same digit so that he could explore several duplicate and different digit placements on the mats. Because Neil controlled the digit placements, he felt more in control of the math *and his learning.*

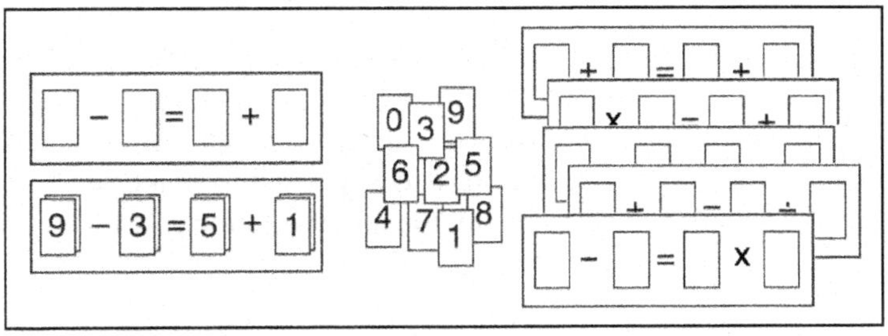

Figure 9.5. Neil's equation mats.

Just as numbers can be composed and decomposed (or put together and then taken apart), equations are also assembled and then disassembled using inverse operations. The trick is to maintain the equal balance even though you are changing, rearranging, and manipulating the symbols. When your child learns how to decompose and compose numbers with number sense, then when the time comes for rearrangement of equations, decomposition and rearrangement are not such strange ideas. It isn't magic, nor whimsy, nor nonsense, and best of all, it is not a secret.

If your child, or you, really needs to have rules in mathematics, then here are two rules to consider instead of trying to remember all of the other rules rattling around in your head.

1. You cannot arbitrarily change the value of a number written within an equation . . . but you *can* change what it looks like to use an equivalent expression so that you can work with it.
2. You cannot change the balance of an equation . . . but you *can* change what it looks like with an equivalent form so that you can do something with it.

Prove It to Yourself:

Make your own equation mats like the ones in Figure 9.5. Use a deck of single digit cards made from cardboard or use the single digit cards from several decks of playing cards. Try several combinations for the equation mats and make a record of the equations that balance. If you are feeling brave, include some fractions and decimals in your card deck.

Figure 9.6. Prove It to Yourself.

KEEP IN MIND

Balance is the key; it is the secret to solving equations at any grade level. The operations allow the composition and decompositions of numbers. Having flexibility in writing those expressions makes solving the more complex problems more manageable. Math isn't supposed to be easy, but it *is* supposed to make sense. The predictability of function equations makes the mathematics reliable as well as useful.

NOTES

1. Hersh, *What Is Mathematics, Really?* 275.
2. de Bono, *New Think*, 125.
3. Liping, *Knowing and Teaching Elementary Mathematics*, 111.

Part IV

CONNECTIONS

> When students can connect mathematical ideas, their understanding is deeper and more lasting.[1]
>
> —National Council of Teachers of Mathematics

Dana, an adult student, described her early experiences in learning math. Her kindergarten experiences taught her to not ask questions— she still remembers those judgmental "behavior adjustment stares" from her teacher. Dana was a curious child who really believed that there was a larger, even possibly spiritual, reason for math and that there were larger meanings for those (in her mind) mystical symbols. She asked questions about what these larger meanings could be only to receive dismissive reactions, possibly because no one really understood why she was thinking this way.

Her wondering (and wandering) mind captured her attention, earning her the reputation of being an inattentive daydreamer. In subsequent years, her daydreaming "worsened," until she finally gave up and accepted that no one could explain an *ultimate rationale* for math to her. No one could explain the "big ideas" of the math. She settled in on a "do or die" survival strategy. Her tactic of choice was to focus *only* on the problem at hand *at the moment*, "do" the work, and ignore anything either before or after the immediate topic or problem.

Eventually, as life would have it, Dana entered college and enrolled in a math course that included trigonometry. She was older, so she had a different perspective and motivation in her approach to learning math. Her curiosity was piqued, so she looked again for those math rationales for "big ideas. " As a result of her searching, she signed up for a calculus course—and even more calculus courses. She found her rationale, and it wasn't mystical; nor was it frowned upon. She had figured out that the mystery was not a mystery. She learned that the explanations she needed were "imbedded in the equations" (her words).

Dana's story is just one variation on the many stories of students who have not learned how to connect math topics or *that there was a connection* between them. The chapters in this section respond to these varied stories that have a central "why" theme by using visual explanations of how some of the math topics are connected. Chapter

10 describes how an ancient area design applies to all those numbers in math class. Chapter 11 is about ratios, proportions, and rate of change and their connections to each other and to other topics. Chapter 12 brings the arithmetic from earlier years into generalized algebraic expressions.

These connections are described with pictures, diagrams, classification comparisons, analogies, metaphors, and stories, because these methods are more effective for the visual-spatial learner. Rote memorizations and isolated facts are not connection-friendly strategies. These same descriptions can also enrich auditory-sequential learners' experiences by offering a deeper understanding of the information.

Children initially encounter the idea behind the rate of change concept as early as the first grade with place value for whole numbers. They see the concept reappear throughout the math curriculum in multiplication tables, the equivalent fractions required for adding and subtracting, percentage calculations, slopes of lines, areas, and as differentials in calculus, among other things. You get the idea—like the Energizer bunny keeps on going, so does the rate of change concept keeps on reappearing throughout mathematics.

You will read about Phil, a middle school student, who successfully completed the linear equation sections in his algebra class, but was struggling with distance, rate, and time problems in science. He was treating the two topics as separate because they were in different lessons, different books, and different classes. Dana, Phil, and all learners can benefit from connecting mathematical ideas because like from the opening quote, "their understanding [will be] deeper and more lasting," their test scores will rise, more opportunities will be available to them, and, best of all, the seeming unsettling, mystical, and secretive unknowns of math will no longer be hidden.

NOTE

1. National Council of Teachers of Mathematics, *Curriculum and Evaluation Standards for School Mathematics*, 64.

Chapter Ten

A Multiplication Area Image for the Ages

Einstein sometimes referred to the source of his ideas as "playing with images."[1]

—Thomas G. West

Einstein, like Theon (35 BCE), had a visual sense of mathematical relationships. Theon's design for calculating square roots is shown on the left in figure 10.1. This arrangement has been replicated over the centuries to illustrate multiplying with different kinds of numbers, taking the application to new heights beyond squares and square roots. The Δx symbol in the drawing means that the sides of the blocks can grow or shrink.[2] Twenty-two centuries later, the twentieth-century Zoltan Dienes's base ten blocks from chapter 5 use Theon's design to illustrate two-digit multiplication shown on the right in figure 10.1.

In this chapter, you will see how Theon's multiplication design and today's base ten arrangements for the multiplication of numbers is applicable to several different kinds of numbers. Multiplication represents the putting together of *same size groups*, a definition that looks and sounds nice and easy with the small whole number amounts, but multiplying with any other kind of number (such as fractions, decimals, or integers) can make grown people mumble and young children cry. In this chapter, the

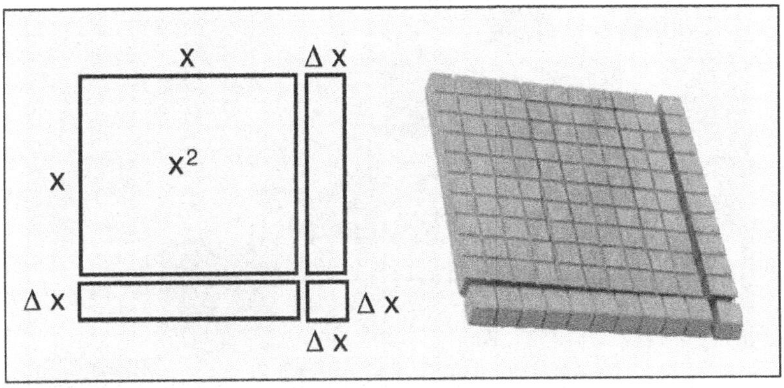

Figure 10.1. Theon's design and Zoltan Dienes's blocks.

rectangular area model serves as a frame of reference to recognize how and why the process works so nicely with those different number types.

FROM THEON TO DIENES: A BRIDGE ACROSS THE CENTURIES

Multiplication with single-digit multiplication "facts," fractions, mixed numbers, positive integers (negative integers, too, but with adjustments), and multidigit numbers all fit Theon's design. For those who need or want the speed of memorizing the "multiplication facts" for immediate recall, go for it. However, recognizing the mathematical relationships—how these multiplication "facts" relate to each other—is worth the effort, especially since the design relationship keeps reappearing.

One example of a fifteenth-century version of a hand-held calculator for multiplication[3] looks a lot like the multiplication table model of wooden blocks developed by a twentieth-century Argentinean math educator, Pastora Sofia Nogues Acuña. A photo of Acuña's wooden table model appears in figure 10.2 (each block in her table

x	1	2	3	4	5	6	7	8	9	10
1	1	2	3	4	5	6	7	8	9	10
2	2	4	6	8	10	12	14	16	18	20
3	3	6	9	12	15	18	21	24	27	30
4	4	8	12	16	20	24	28	32	36	40
5	5	10	15	20	25	30	35	40	45	50
6	6	12	18	24	30	36	42	48	54	60
7	7	14	21	28	35	42	49	56	63	70
8	8	16	24	32	40	48	56	64	72	80
9	9	18	27	36	45	54	63	72	81	90
10	10	20	30	40	50	60	70	80	90	100

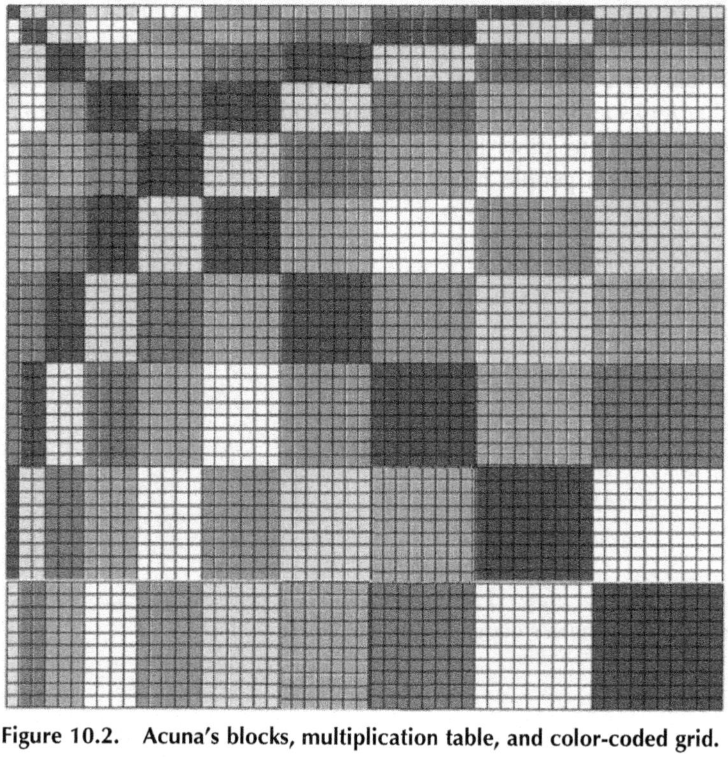

Figure 10.2. Acuna's blocks, multiplication table, and color-coded grid.

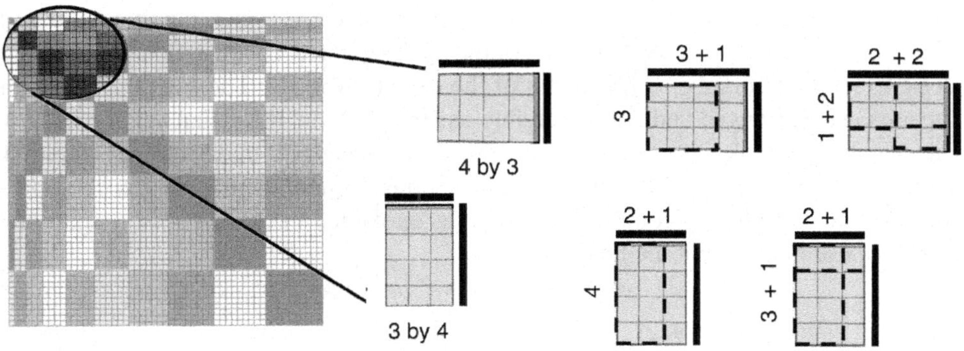

Figure 10.3. Enlargement for 3 × 4 with composition designs.

represents a "multiplication fact") and is placed above a shaded grid of the same multiplication table. The rectangles in Acuña's multiplication blocks table are some of the same spatial number personalities cut out by the fourth graders in chapter 3.

The side measures of the rectangular blocks come from the rows and columns in the grid. For example, select the block that is in the third column (starting from the top left) and move down that same column to the fourth block. The width is usually written first, so the block is 3 units wide and the length is 4 units long. The related multiplication "fact" is 3 × 4 or 4 × 3 (same area just rotated). The multiplication "answer" ("fact") is represented by the area of the rectangular block. The shapes in figure 10.3 show the 3 × 4 block alongside the 4 × 3 block and several flexible decompositions and compositions of the blocks marked with dashed lines.

SAME DESIGN FOR MIXED NUMBERS, FRACTIONS, AND DECIMALS

That 3 × 4 wooden block in figure 10.3 will increase in size when fractions are tacked on to the sides as it grows in the same way that the Δx shows a change in Theon's design in figure 10.1. That newly increased area arrangement for multiplication could look like the arrangement in figure 10.4 with the addition of a ¼ strip to the 3 side and a ½ strip to the 4 side. With the tacked on fraction measures to the 3 × 4 block, the new multiplication design reflects the multiplication of mixed numbers.

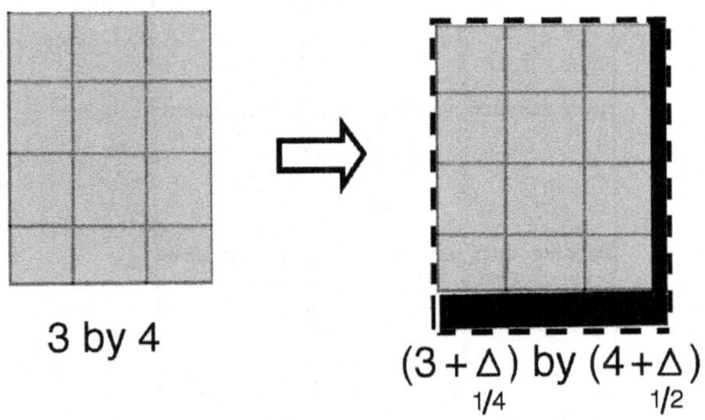

Figure 10.4. The 3 × 4 block showing how the sides grow.

The design in figure 10.5 shows multiplication of these mixed numbers, 3¼ and 4½, and is enlarged to show the details of the newly created Theon area multiplication design for mixed numbers. The width of the design in is 3¼. The fraction blocks in the area are rearranged and composed into squares to show the visual version of how the calculations look with shapes. Four of those ¼ pieces will make another whole "1 unit" square, and two of the ½ pieces make another "1 unit" whole square. The addition of ⅛ and ½ blocks requires like terms (as explained in chapter 4). Compare the area of these blocks with the total 14⅝.

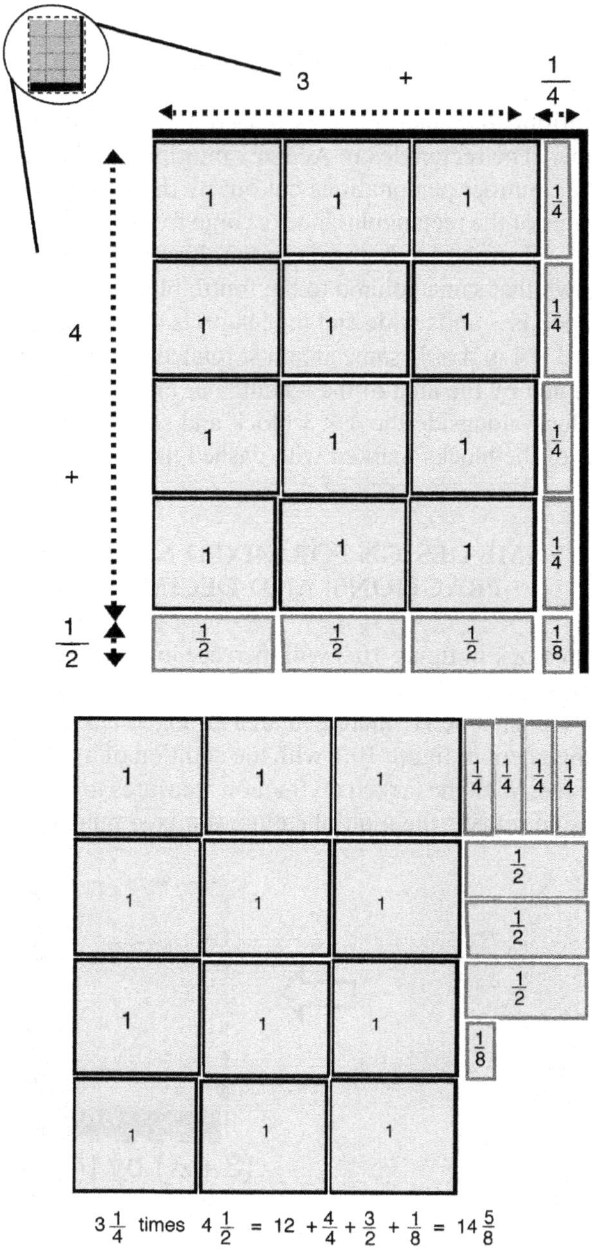

Figure 10.5. Enlargement of 3 x 4 block to show multiplication of mixed numbers with Theon's design.

One unit enlargement of the 1 x 1 block in the multiplication table.	A picture of four pieces each representing fourths of the 1x1 block.	A picture of three pieces each representing thirds of the 1x1 block.
	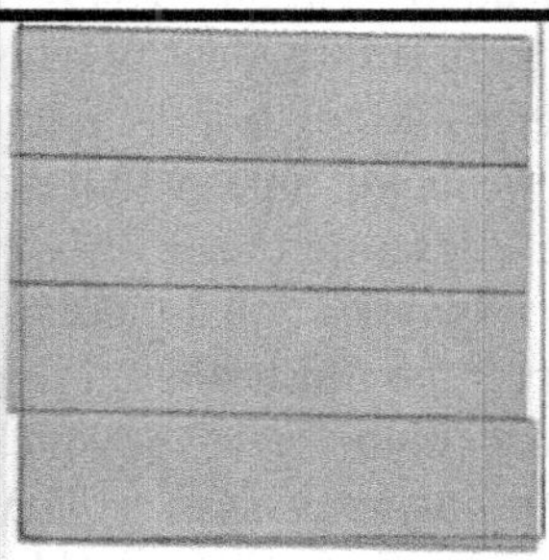	

Figure 10.6. Fraction tile blocks.

Theon's area model also applies to fraction multiplication, but the fractions need to be decomposed from the unit square in the far upper left of Acuña's multiplication table in figure 10.2. That single "1 unit" square area block is enlarged in figure 10.6 and is located on the left. The next two sections show different ways to represent a square unit piece: one section has fourths (just like the ¼ areas in figure 10.5), and the other section shows thirds.

To show how these blocks illustrate multiplying with fractions requires that each fraction *area* must be distinguished from the fraction piece *edge*. To find area, multiply length edge times width edge. Distinguishing between sides and area is not always an easy task, even with whole numbers, but the difficulty is compounded when using fractions. Ned, a fifth-grade student, was confusing the side measures with the area measures. To Ned, the one-fourth piece of the one unit represented the area *and* the side. Ned learned to roll the fraction piece to make a track (perimeter edge) of the sides so that he could "see" the line side as different from the area.

Ned used the fraction tile (an area block piece) for ¼ and placed it on a line to roll to mark the sides. He marked the end of the "1" side then rolled the tile to the right and marked where that side ¼ side ended, as shown in figure 10.7. He only needed two adjacent sides to multiply for length times width, so he labeled each side with

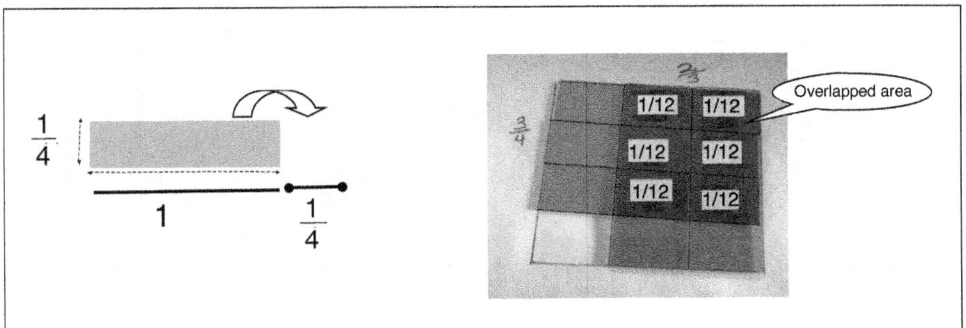

Figure 10.7. Area to side by rolling the block and showing multiplication with fractions using Theon's design.

the correct measures, "1" and "¼." The rolled lengths measure the *linear sides* so that the "length times width" multiplication is used to figure out the inside fractional area of the tile piece. Both the area and a side have the "¼" number; no wonder this is confusing to a child!

The roll of the one-fourth piece helped Ned see the one-fourth edge for multiplying with fractions. He could then imagine placing three one-fourth tile pieces to get the ¾ that he needed for the multiplication example of ¾ × ⅔ shown on the right in figure 10.7. Next, he did the same kind of roll to identify the sides of the ⅔ fraction pieces. The *edges* of the ¾ side and the ⅔ side are now the fraction numbers in the multiplication problem. These edges are the length times width that your child already knows as an area calculation. Where these fraction tiles overlap shows the resulting "answer" area for the multiplication of the two fractional pieces.

There are two ways to interpret the illustration in figure 10.7. One is numeric with number symbols (sometimes called algebraic, in which is discussed in chapter 12), and the other is geometric with geometry shapes. The number version would be ¾ × ⅔ = 6/12 = ½. The geometry version would count the 1/12 pieces in the overlapped section and then count all of the 1/12 pieces in the full unit square. The number of the overlapped pieces is the numerator, and the total number of pieces is in the denominator. Both versions show the same relationship.

Yes, this 6/12 ratio fraction (six 1/12 pieces in the overlap is in a ratio with all twelve of the 1/12 pieces in the total) will reduce to its equivalent ½ ratio. If two of the 1/12 pieces are moved, rearranged, and placed next to the ¾ side, then the ½ equivalence section of the whole area is easier to recognize visually. Now all six of the 1/12 sections are on the same "half side" of the total area and actually look like half of the total area.

The mixed number diagram in figure 10.4 includes a multiplication of fractions in the lower right-hand corner. In figure 10.8, that lower right-hand corner has been enlarged to first see the corner of the mixed number problem, and then enlarged again

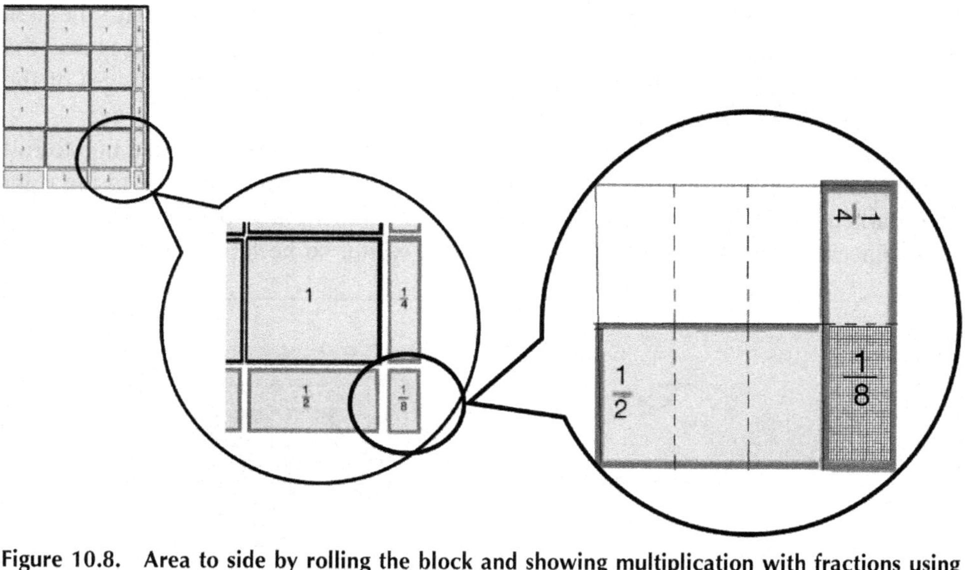

Figure 10.8. Area to side by rolling the block and showing multiplication with fractions using Theon's design.

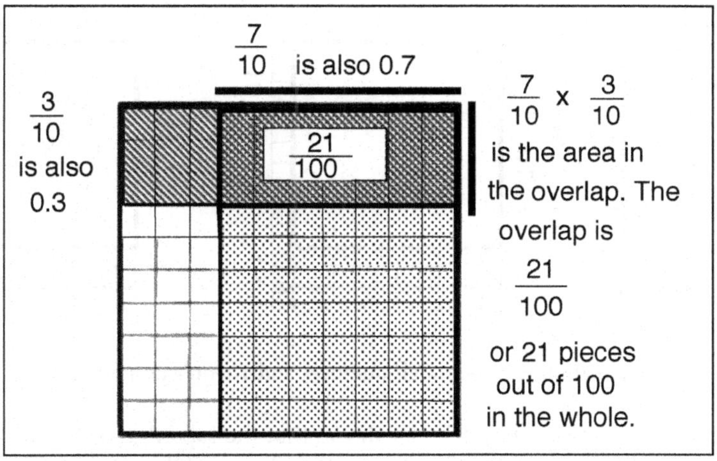

Figure 10.9. Multiplication with decimals also has Theon's design.

to focus on the fraction multiplication piece of the multiplication of ½ × ¼ *with* overlapping sections. In the second enlargement, the ½ side (rolled by Ned) of the ½ tile (area) piece is multiplied by the ¼ side (another rolled side) of the ¼ (area) fraction piece. The result is the ⅛ overlapped area piece. Keeping the area measures distinguished from the sides helped Ned "see" the relationship.

Multiplication with decimals also fits Theon's area arrangement that was used with fractions. The fractions will have denominators with tens, hundreds, or multiples of tens, based on the place value description in chapter 4. The picture in figure 10.9 represents $^7/_{10} \times {}^3/_{10}$ (the two adjacent sides) and the part that is covered by both fractional pieces represents the overlapped area of $^{21}/_{100}$ (there are twenty-one pieces in the overlap section and one hundred same-sized pieces covering the whole unit). Again, there is a number version and a geometry version for understanding how the overlapped area shows the multiplication.

Whether you are using Theon's ancient design, Sawyer's 1964 illustration, McLean's 1972 blocks, Mary Laycock's 1987 pictures, or the upgraded 1990s and 2000s versions of the block area, the same arrangement works. This representation is really an old idea that just keeps coming around again, a reminder that real mathematics has staying power beyond "the next test." Good designs, like good strategies, have staying power.

VISUAL-SPATIAL MULTIPLE-DIGIT MULTIPLICATION

Following the lead of the other visual Math Aficionados, Lola June May described in a 1991 article[4] how a visual breakdown of how the Dienes's base ten blocks could be used to explain a two-digit multiplication problem. The area design 11 × 13 arrangement in figure 10.10 was adapted from May's article and shows how the numbers for the sides and top are decomposed, just like the design example in figure 10.3. Because the design is rectangular, the same measures match the opposite sides of the rectangle.

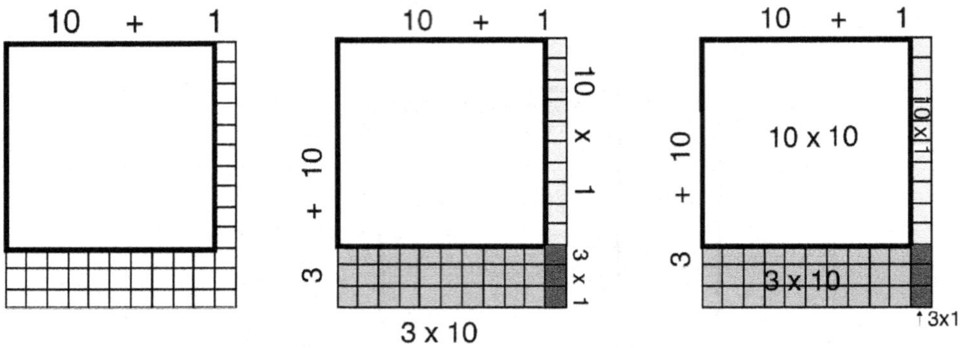

Figure 10.10. Multiplication with two-digit numbers and Theon's design.

The base ten diagram in figure 10.10 shows that the top edge of the arrangement is 10 + 1 (or 1 + 10, if reading in a left-to-right direction) and the side edge is 10 + 3 (or 3 + 10 if reading bottom to top). This kind of area arrangement will return in algebra and is described in chapter 12. Visual-spatial learners can finally *see* the multiplication design and, like James and Jake, can make sense of the digit locations and no longer have to memorize "digit placements" for all those two-digit multiplication practice problems. The total area is the total count of all of the pieces. The digit placement clues are in the area—just match the block sizes.

KEEP IN MIND

Deeper understanding always lasts beyond the next test. Without a lot of time taken up with relearning or re-explaining, there is more time available to make more connections with new knowledge. Deeper understanding provides inherent clues for retrieving any temporarily lost pieces of information caused by lapsed time or brain hiccups. Besides the benefit of remembering, the information retrieved will probably make sense *and* be correct! Talk about coming around again for connections—from Theon to third-grade multiplication and then on to an algebra classroom!

> The simplest explanation of a visual-spatial learner is that they generally think in pictures, rather than in words. They also tend to learn holistically, instead of sequentially, or in parts. The visual-spatial learner can easily see the big picture of things, but might miss out on the details.[5]

—Linda Silverman

NOTES

1. West, *In the Mind's Eye*, 41.
2. Hogben, *Mathematics for the Millions*, 278.
3. Hogben, *Mathematics for the Millions*, 89.
4. May, "Developing Multiplication Concepts," 447.
5. Silverman, *Upside Down Brilliance*, https://www.time4learning.com/visual-spatial-learners.shtml.

Chapter Eleven

Ratios, Proportions, and Rate of Change

All is connected . . . no one thing can change by itself.[1]

—Paul Hawken

Rarely does any one thing change in isolation. Measuring and comparing those interdependent changes is what rate of change is about. The concept starts very early in first-grade place value lessons when the columns represent increasing by tens. In the early grades, ratios are first expressed as fractions; later the fraction format is used to express a ratio that *compares* amounts in a variety of ways and in a variety of contexts. These ratios are then written as proportions. Making decisions *about the comparisons of these change measurements* is the mathematics of managing rates of change.

Because the written proportions can have missing numbers in four different locations, depending on the problem being solved, they seem to be filled with clutter and "moving targets" for the Math Avoider, and neither clutter nor moving targets is very conducive to stability. The number-clutter caused by the different contexts and the changes in the expressions are what plagued a middle school student named David.

David had passed all of his tests related to fraction skills, yet was confused when confronted with rate of change and proportions. David had dutifully collected these concepts as he progressed through the grades, but he had not connected the topics in a meaningful way. This chapter's goal is to explain how David tackled his number-clutter and controlled those moving target elements in rate of change. David's experience can help you declutter and clarify any mixed messages that confront your child or that you may have endured along the way. The clutter gets cleared and the "moving targets" in proportions become predictable and manageable.

For examples of some "rate" concerns, think about what a racecar driver, a marathon runner, and a speeding ticket holder have in common. All have experienced speed as a rate of change between distance and time, either as a benefit or as a consequence. The driver and the runner are conscious of the distance to time ("d/t") ratio relationship because they consciously time their laps or mile markers, so that they can manage their respective rate of speed to achieve their goal. The speeder is probably just impatient or running late, so the speeding ticket becomes their "managing agent." All are aware of the rate of speed, just at different levels.

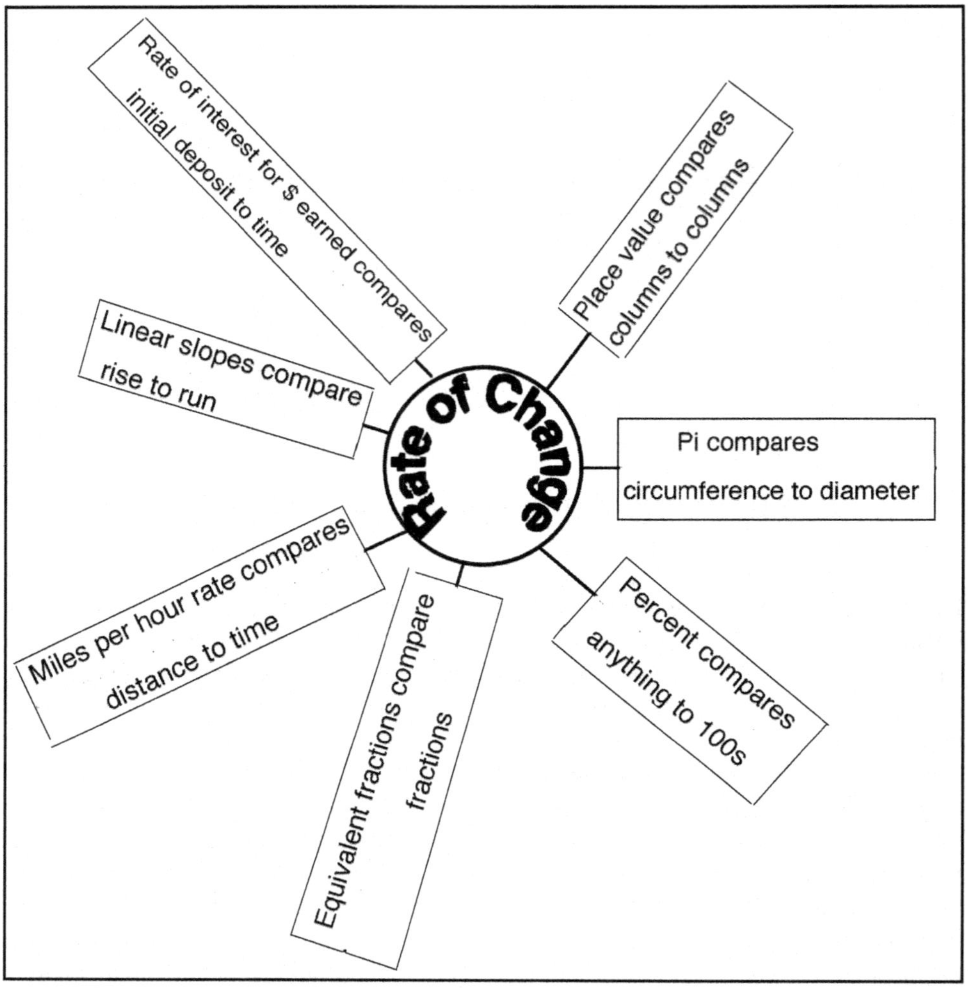

Figure 11.1. Concept web for rate of change examples.

Some of the proportion comparisons confronting David, and others facing number-clutter and moving targets in proportions, are displayed in figure 11.1. A few of these examples are examined in this chapter, and the others are left for you to explore after you understand the organization and management of the predictions, or better, how to control both. Organizing these proportion examples first will interrupt clutter blindness so that the proportion contexts no longer are "hiding in the mess" and *not seen at all.*

Fractions in the fourth grade lead to proportions in the sixth grade, and to rate of change in middle school or algebra. Just passing a test on fractions without really understanding that a fraction is both a noun (a number like ¾) and a verb (represents division when 3 is divided by 4) will make proportions and rate of change much more complicated. There are too many "Davids" who pass the tests but do not know that they have any control over how fractions can be used in proportions and rate of change. Fractions become the gatekeepers to that all-important mathematical concept called rate of change. This disruption caused David's number-clutter.

FRACTIONS AS GATEKEEPERS

Too often, fractions are the uninvited gatekeepers to learning about the larger concepts of ratios, proportions, and rate of change. Fractions really represent just one kind of ratio, the kind that compares a particular number of pieces of something to the total number of pieces of that *same "something."* All of the pieces of a fraction belong to the same pie, apple, or other *common category*. Ratios, on the other hand, can represent an active comparison between *two different categories*, not always the same pie or apple. For example, a ratio can compare miles with hours. These are clearly not in the same category, because one is distance and the other is time.

Years and tests are dedicated to learning fractions, finding common denominators, and dividing with fractions. When assessments only focus on adding with common denominators or inverting and multiplying as the way to "divide" fractions, then the message of fractions as a comparison is confused, and the calculation procedures become the message instead of the fraction concept. Somewhere in between lowest common denominator (LCD) and "invert and multiply," the real message is lost—the message that fractions are ratios that show just one kind of comparison of part to whole.

Ratios "obey the same rules" as fractions when they are added, subtracted, multiplied, or divided. That is because the "rules" applied to ratios *first*. They both quack the same and, like ducks, can be identified as "duckworthy" and available for combining and taking apart after they have been "freed from their related objects." Although the comparisons can never be completely freed, the combinations and decompositions only have meaning when the identities of the items being compared are reintroduced after the operations. Keeping faithful to what the comparisons actually compare is what makes the predictions meaningful.

DAVID'S DECLUTTERING FRACTIONS

David's cluttered numbers perspective was filled with many different representations of fraction tiles, pennies on top of the fraction pieces, quarters that represent twenty-five pennies, fractions as part of the one hundred pennies in the dollar, percent notation, a full sheet of paper covered with equivalent fractions to one-fourth. David's immediate challenge was to learn about the percent ratios that preceded a proportion lesson. All of these cluttered his mind and thinking.

David had assumed that the 10 percent way of writing fraction numbers applied to all percents. For example, 10 percent meant that you divided by ten, so 25 percent meant that you divided by twenty-five. When that did not work, he used what he believed to be the next best line of reasoning: if 10 percent of 45 is 4.5, then 25 percent of 83 was 8.3 as well since they are both percents. To David, his responses made sense and were located somewhere in that number-hoarder's house of numbers. It was as if the "stack of 10 percent relationships" had toppled over and mixed in with those other crowded stacks of 25 percent and 50 percent numbers and all the other percents.

David is not alone in his clutter-blindness. He is joined by countless others who have found themselves in the same conundrum even after passing all of those tests, doing all of the earlier homework, inverting-and-multiplying with the best of them, calculating all of the common denominators whether they were needed or not, and even easily making change because he helped out in the family business. Imagining how one representation could morph (change and look different) into another *and still stay the same value in his mind* turned out to be one of the culprits, even though he had passed all of the tests on fractions!

The solution for David was to first clean up the mess *under his definitions of what constituted the mess*. All of those pieces cluttering his mind were separated, sorted, described, and even labeled. Because he had liked and understood the fraction tiles earlier, that was the starting point for building, and rebuilding, *his own understanding* of the connections. During every stage of his "fraction metamorphosis," David described, questioned, made Halmos-like vague guesses, and confirmed his own understanding through several trials and number tests, with a little calculator assistance. Talking about and describing his clutter helped David think!

Fraction strips like the ones in figure 11.2 helped David organize and "control" the equivalence ratio number clutter. He recognized the number patterns in the numerators and denominators of the equivalent fractions from the multiplication table. He could write all the equivalent fractions on any page of homework, but he had not recognized that the rate of change *is the multiplier* from the multiplication table.

If your child challenges the doubling rate because the "$^{6}/_{9}$" is not double the "$^{4}/_{6}$," then your child is asking a good mathematical question. Your child *should* challenge ideas if they do not make sense, and then keep trying to make personal sense of any patterns. Equivalent fractions are calculated by multiplying the numerator and the

$\frac{1}{5}$	$\frac{2}{10}$	$\frac{3}{15}$	$\frac{4}{20}$	$\frac{5}{25}$	$\frac{6}{30}$	$\frac{7}{35}$	$\frac{8}{40}$	$\frac{9}{45}$	$\frac{10}{50}$	$\frac{11}{55}$	$\frac{12}{60}$	...

$\frac{3}{4}$	$\frac{6}{8}$	$\frac{9}{12}$	$\frac{12}{16}$	$\frac{15}{20}$	$\frac{18}{24}$	$\frac{21}{28}$	$\frac{24}{32}$	$\frac{27}{36}$	$\frac{30}{40}$	$\frac{33}{44}$	$\frac{36}{48}$	...

$\frac{2}{3}$	$\frac{4}{6}$	$\frac{6}{9}$	$\frac{8}{12}$	$\frac{10}{15}$	$\frac{12}{18}$	$\frac{14}{21}$	$\frac{16}{24}$	$\frac{18}{27}$	$\frac{20}{30}$	$\frac{22}{33}$	$\frac{24}{36}$	...

$\frac{1}{4}$	$\frac{2}{8}$	$\frac{3}{12}$	$\frac{4}{16}$	$\frac{5}{20}$	$\frac{6}{24}$	$\frac{7}{28}$	$\frac{8}{32}$	$\frac{9}{36}$	$\frac{10}{40}$	$\frac{11}{44}$	$\frac{12}{48}$	...

Figure 11.2. Four different equivalent fraction strips.

Ask Yourself:

How can I figure out the rate of change in a strip if the sequence does not always have doubles?

These three strips have been removed from the 2/3 fraction strip. All of the fractions in these three strips are doubles. How could you put them back in order into one single strip? Where would you put them so that they show a consistent rate of change?

$\frac{2}{3}$	$\frac{4}{6}$	$\frac{8}{12}$	$\frac{16}{24}$	. . .

$\frac{6}{9}$	$\frac{12}{18}$	$\frac{24}{36}$	. . .

$\frac{10}{15}$	$\frac{20}{30}$	. . .

Could you do this with the $\frac{3}{4}$ strip and the $\frac{4}{5}$ strip?

Figure 11.3. Ask Yourself.

denominator *by the same* number. Said another way, the rate of change for the numerator must be consistent with the rate of change in the denominator.

CHANGING NUMBERS, CHANGING LOCATIONS, AND MOVING TARGETS IN PROPORTIONS

Hour totals change as mileage changes (miles per hour), money amounts change with different interest rates (interest times deposit), number of heartbeats change per minute (bpm) with different stresses, and a circle's circumference increases as the diameter increases (pi or 3.14159 . . .). Each member of a pair of measurements *changes "in partnership" with the other for a given rate of change*. These "partners" can be measured and compared to calculate a desired and predictable rate of change.

Proportions and rates of change can seem to have moving targets, as the *locations* of the numbers and the missing pieces move around in the proportions or the rates themselves change in a comparison shown later in figure 11.7. These are only two

Figure 11.4. Cuisenaire Rods.

examples that make for an unsteady foundation when trying to "solve proportions." Cross-multiplying isn't the answer to this moving target dilemma because it only works in specific circumstances, such as when the two ratios are equal. Making sense of the overall proportional equivalence concept is a preferred option because that doesn't change even when number locations or rates change.

For the visual learner, having a consistent diagram that fits the larger concept can help. Long proportional rods, such as the Cuisenaire Rods shown in figure 11.4, have helped students understand the comparisons of lengths and related number relationships. These rods are initially used in the early first, second, and third grades to show addition and multiplication. Later, the same rod alignment can illustrate fractions and other mathematic symbolic expressions. For example, the longer arrangement of rods in figure 11.5 shows $4 \times 5 = 20$, ¼ of $20 = 5$, $5 + 5 + 5 + 5 = 20$, $20 \div 4 = 5$, and 25 percent of 20 is 5—one arrangement with five, and possibly more, interpretations.

The earliest exposure that children have to rate of change happens subtly in that first-grade place value notation, possibly another reason why place value is so difficult for young children to understand. The *same digit* can move to *another location* and still be the same digit, but now it means something that is ten times larger. To young Kendra, that digit just moved to another location. It still looks the same, so why do they say it changed? Although multiplication isn't introduced until two years later, your child and Kendra are expected to understand this multiplier rate-of-change idea when they move digits in place value. Moving target, indeed.

The rate of change between the place value columns is essential to the organization of a "times ten" rate of change increase. A diagram of the base ten rods (or only the

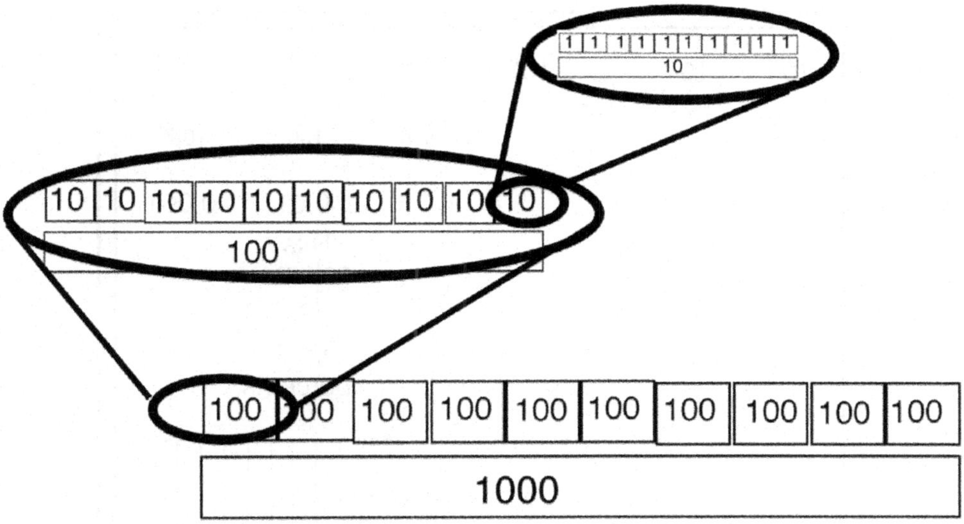

Figure 11.5. Place value enlargements showing the "10" rate of change on whole number side of decimal point.

tens and ones from the Cuisenaire Rods) can show the tens relationship arrangements of how the tens groupings "grow" and affect the place value ratio rate of change by tens columns. The groups of ten repeat the same ten-to-one ratios with changes to the amount within each block. The diagram in figure 11.5 shows visually how this same ten-to-one repeats, yet the relationship continually makes larger numbers, a very complicated idea for very young minds.

Place value and percentage problems are very closely related because of the "tenness" in both systems. Percentages are related to the "one hundred cents in a dollar" and are always compared with hundredths. In other words, whatever is being compared must be divided into one hundred pieces, just as it takes one hundred cents to equal one dollar. The illustration in figure 11.6 shows how the hundredths location in place value is associated with percents, and also shows an expansion enlargement perspective with decimals similar to the "growth by tens" with the place value of whole numbers in figure 11.5. Both examples show a repeated expansion (or reduction) by tens.

Percentage problems can seem to have those "moving targets" in proportions because the "x" keeps moving to different locations, depending on the missing piece of information. Several typical examples of these percentage comparisons are illustrated in figure 11.7 along with different ways to express the same picture but with different missing pieces and the changing location for the "x unknown amount." David did not know that he had "permission" to change the numbers or "move the x" in proportions, *as long as he maintained the same ratio value*. Flexible thinking was not in David's math past, but it is essential for your child's math future.

Some Math Aficionados say that a percentage is a number; others disagree and say that a percentage is a rate. Your child, and maybe you as well, can find it difficult to tell the difference because the technique of "changing the percentage expression into a deci-

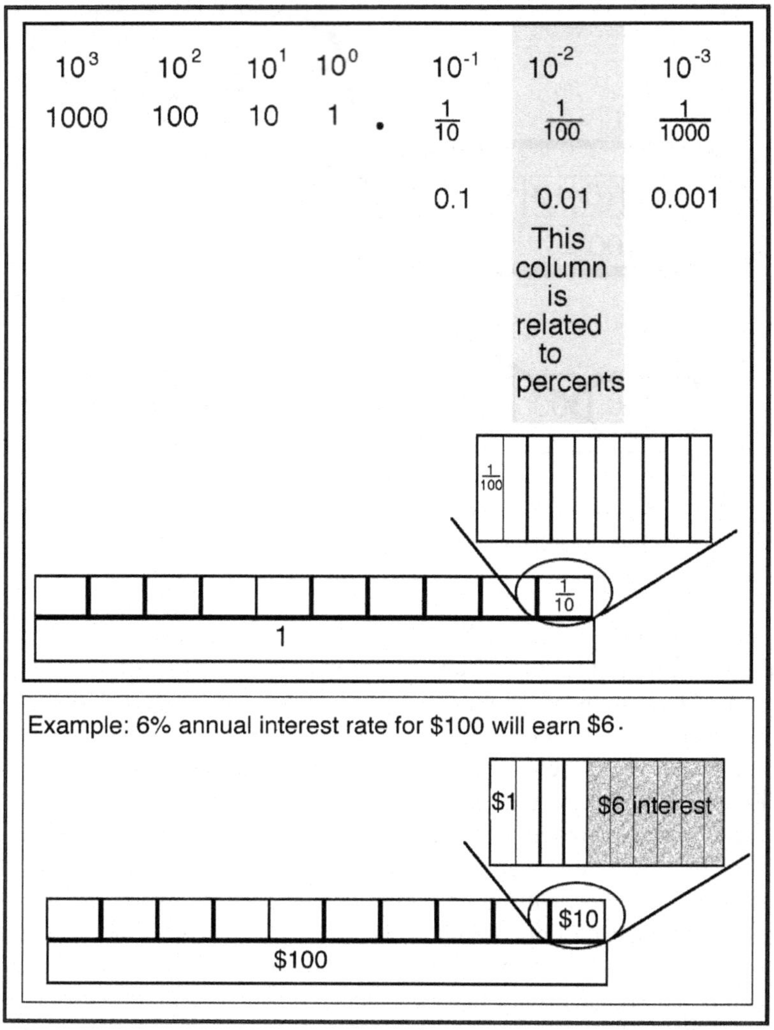

Figure 11.6. Place value enlargements showing the "10" rate of change on decimal number side of decimal point.

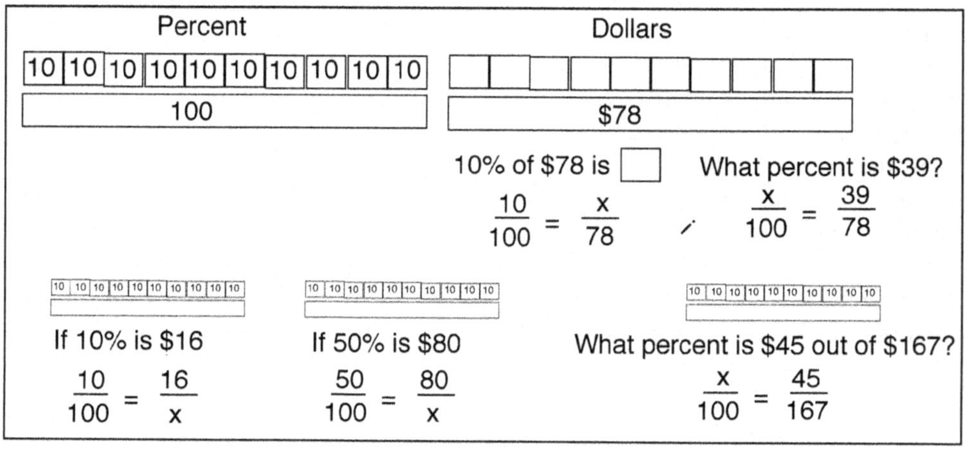

Figure 11.7. "Moving targets" for proportions—missing numbers in different locations.

mal number and then multiplying" seems to fluctuate back and forth with the number and the ratio. It certainly looks like a number when that percentage changes to a decimal! The transition happens in the place value location *when the numbers get compared.*

Ratios measure "what's going on." Some comparisons make sense while others do not, so *the choice* of what to use in a ratio depends on a dynamic that you choose to measure and predict. The others get tossed. For example, those motion word problems beginning with "a train leaves the station" that are in math phobia nightmares are just about distance covered, regardless of where they start. Whether operating a train, driving a car, or painting a room, the action is happening *over time*. The ratio is about measuring the distance traveled (or the painting work completed) and comparing that distance to a measurable *amount of time*.

The prediction is about how fast you can travel, drive, or paint or what your rate of activity must be to finish in a specific amount of time. Other ratios happen when buying items. While the context and the proportional amounts will vary, the "meat and potatoes" of the mathematical relationship based in a rate of change does not change. Managing the prediction lies in how you control your driving rate in miles per hour or select a download speed in Mb per second (or higher). These rates-of-change proportions, and others, can be illustrated in rod lengths like the ones in figure 11.8.

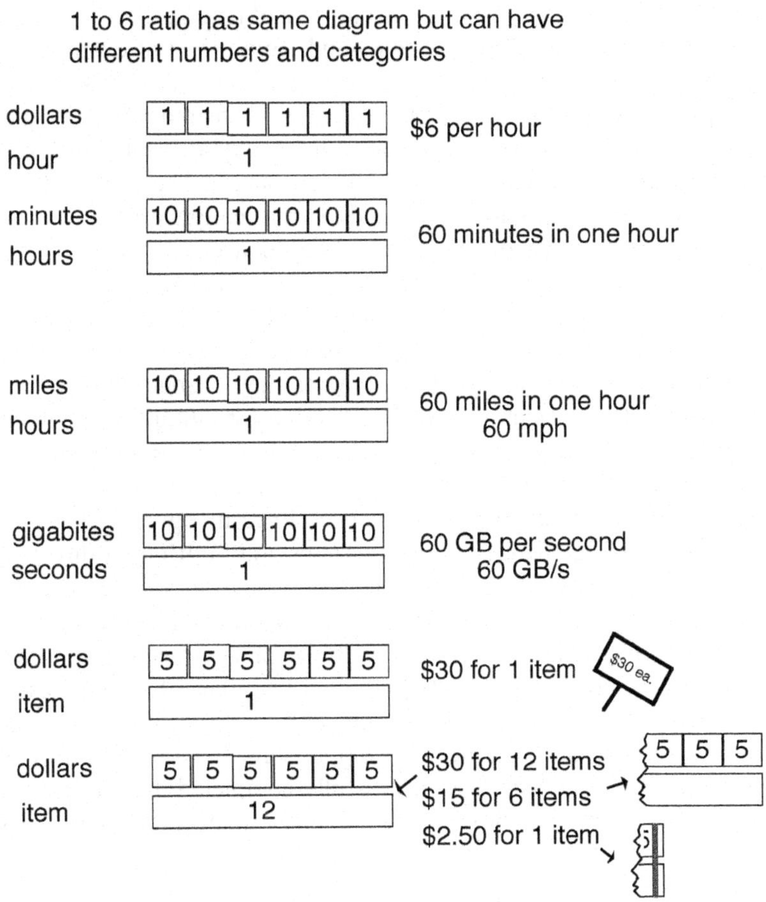

Figure 11.8. Different examples for 6 to 1 ratio.

This information helps you help your child because your language changes from a "how to do it" ritual to an explanation grounded in a consistent and connected way. For example, when talking about percentages, instead of saying something like, "Change the percent to a decimal and multiply," suggest looking for connections. "Look for the percent number in the place value mat," or "look for two equivalent fractions where one hundred is in the denominator for one of the equivalent fractions in a strip," or "look for hundreds." Even a comment like, "I don't know, but let's find it together," is better than, "Just do it."

SLOPE AS A RATE OF CHANGE

When your child enters a pre-algebra or an algebra class, they will again encounter rates of change, this time with linear equations. Linear equations are generalized expressions of how data changes in a constant way; the data (the collected information for the pattern to "discover itself") is recorded in a table format with "x" and "y" axes. The ratio that comes from comparing the change in "y" with the change in the partnered "x" from the data tables is called the slope ratio. A high school student named Ella dubbed this ratio as the "go-to idea" that will connect any math examples that involve measured changes that have a constant rate of change.

Ella's idea was about describing, in her mind, how different examples for ratios, such as miles per hour, beats per minute, megabytes per second, risers compared to treads on steps,[2] or other constant change rates could fit into a linear equation. She was also referring to all the other times when a linear equation relationship appears in math, science, economics, chemistry, and physics courses. The concept keeps repeating, so she might as well think through it and learn it really well.

Math Aficionados say that the slope of a line "is a ratio of the change in the y coordinate (symbol for that change notation is Δy) compared to the change of the x coordinate (the symbol for that change notation is Δx). This means that both change in "partnership" or, said another way, a change in one makes the other one change. You can express it any way you wish, as long as they change together *at the same rate*. Remember, that Δx was used in Theon's design for multiplication in chapter 10, another reminder that clearly thinking through the math the first time will make any information, new or old, easier to connect to other topics.

The linear equation cards[3] from the *Algebra Game* in figure 11.9 have different slope ratios and therefore change at different rates and in different directions. These different slope ratios come from comparing "the change in x with the change in y" *for each separate line*. For example, the "information of distance" changes in relation to its partnered "time," whether driving, walking, or late arriving somewhere. The "information of download time" to its partnered "number of megabytes" will give information about which online service to subscribe to. These slope rate relationships can also be illustrated with the earlier rod comparisons, like the ones in figure 11.5.

Phil, a middle school student, had successfully completed the linear equation sections in his math class and was now struggling with distance, rate, and time problems

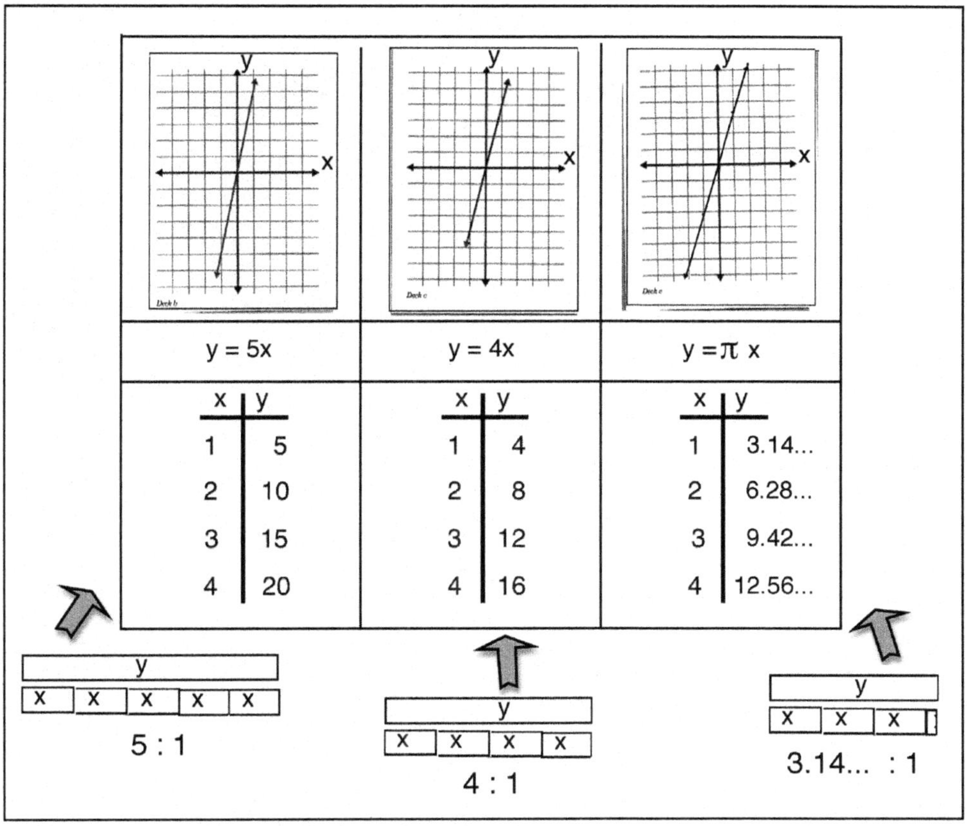

Figure 11.9. Linear slopes with algebra cards and rods.

in science. He knew the slope ratio and he was familiar with the idea that cars move at different rates. However, like other students, he considered the two topics to be separate because they were in different lessons *and* in different books. After the linear equation (y = mx + b) was placed next to the distance, rate, time equation (D = rt), Phil saw that the "m" lined up with the "r," the "x" lined up with the "t," and the "y" lined up with "D." He recognized the connection and was delighted!

KEEP IN MIND

Children learn early how to cope with math. David had learned how to survive the same way Dana had described her own kindergarten experience, by not asking questions and instead "working" each math problem individually in order to get through each day. Robbie had learned that "you get in less trouble if you answer questions quickly; the faster the better." Phil didn't know that he could look for connections across topics. A change in your perception about ratios, fractions, proportions, and math in general will help you help your child by *changing your language so that they learn better ways to cope.*

NOTES

1. https://www.brainyquote.com/quotes/p/paulhawken382086.html
2. Draper, *Winning the Math Homework Challenge*, 121.
3. Draper, *The Algebra Game*, deck c.

Chapter Twelve

Algebraic Thinking

> Common arithmetic and algebra rest on the same computational foundations.[1]
>
> —Sir Isaac Newton

Algebra is about thinking. Although the language is symbolic, the thinking is strategic. Suggesting that algebra isn't necessary is like saying that thinking isn't needed. If you agree with Isaac Newton that algebra generalizes arithmetic and that they both rest on the same foundation, then most of the algebraic thinking work has already been completed in the earlier arithmetic work. Now the task is to "graduate" the relationships among numbers in arithmetic to letter symbols in algebra and express those relationships in a kind of shorthand called equations. This chapter is about how these numbers graduate into letters, yet still share the same foundation.

THE SHAPES FOR AN ALGEBRA TRANSITION

The base ten blocks in figure 12.1 represent the first three shape pieces of place value from chapter 5. The same shapes represent the three pieces of a quadratic equation used in chapter 7 and chapter 8 with patterns and in chapter 10 with areas. Now it is algebra's turn. The large square is named "x squared" (x^2) because the actual size is unknown, however we do know it is a square. The stick-like shape has the same length as the side of x^2 so it has an area of "x" (x length times 1 unit width). The small square is a 1×1 square. Think of "x" as "I don't know what the number is yet."

Since the shapes are the same, it would be reasonable to suggest that the arrangements can also be the same, whether you are using the Dienes's base ten blocks or algebra tiles. The edges of both block designs are read the same way. The left picture example of blocks in figure 12.1 is read as $(10 + 3)(10 + 1)$ and the right example is read the same way, replacing the 10 with an x to read $(x + 3)(x + 1)$. When your child looks at these numbers, *really looks at how the blocks are arranged* like they were in chapter 10, then the two arrangements turn out to not be all that different. Both arrangements are written as binomials (two terms).

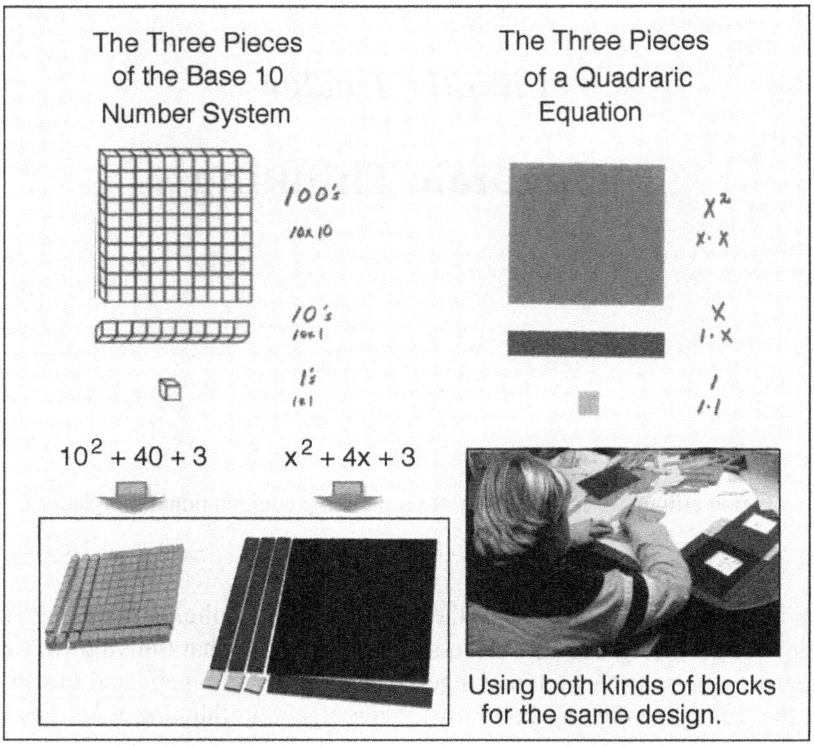

Figure 12.1. Three pieces for arithmetic and three pieces for algebra.

Math Aficionados may choose to use more letters for more variables and assign an area for the smaller square as "y squared" with a side length "y" instead of 1 unit. The number isn't as important as the relationship, so, for simplicity, start out with only one variable at a time to name the blocks. In order to "solve" (figure out what is missing) for those missing numbers in these relationships, a specific relationship needs to be identified—either equals, less than, or greater than. Again, to ease into the generalizations, let's use our soulful "equals" balance and allow that area to change as the sides change.

This algebraic generalization for number relationships is accomplished through letters (Math Aficionados call them variables) that take turns *representing numbers* in equations or other symbolic relationships. Because letters are now replacing numbers, that multiplication "×" symbol needs to be retired from meaning multiplication. It's just easier and less troublesome to not have one symbol with two interpretations. Instead, the symbol for multiplication is revised and written as a centered dot (•).

First, during the early grades, numbers are those abstracted representations of things, or "freed from apples to become integers."[2] This is the initial level of generalizations. Next, in the upper grades, number relationships are studied and described in a second level of generalization. Then finally in the algebra years, those number relationships are expressed in a letter version for generalized expressions that will *embrace all numbers* (unless otherwise restricted). It's the *thinking* that makes all of this happen.

FROM MULTIPLICATION TABLES TO COORDINATE TABLES

The multiplication and addition tables in chapter 7 produced quite a few patterns, and each of those patterns had an expressed relationship. The tasks in chapter 7 and chapter 8 were to identify the patterns and express them by how they were composed. The tasks in algebra are to symbolically express the relationships with an equation and then plot them on coordinate graph axes. The diagram pictures (graphs) of number relationships provide an image of what the relationship looks like. Figure 12.2 takes one of the examples from chapter 7 and shows how that relationship transforms into an equation and its related graph diagram.

Whether your child "saw" the numbers as square numbers or square arrangements, the pattern message was the same. The mathematical thinking started happening when your child *thought about* how to identify the pattern. The mathematics continued when the pattern was described, then written as an equation, and later graphed. The table in the algebra format *generalized the trend* of the squared numbers and allowed the pattern equation to include negative integers. That solid graph curve implies that the trend also includes all of those numbers in between the integers—the fraction ratios and the mixed numbers.

Another example from chapter 7 that shows how the shapes graduate from an addition table to an algebraic table is shown in figure 12.3, where the algebra equation can be expressed with multiplication or with addition, *and still express the same relationship*. The expression "2x" means that the "number that replaces x" is doubled (multiplied by 2 so there are two "xs"). The expression "x + x" also means that there are two "x numbers" that will be added, but you don't know what those numbers are, *yet*. In both cases there are two numbers represented with an "x."

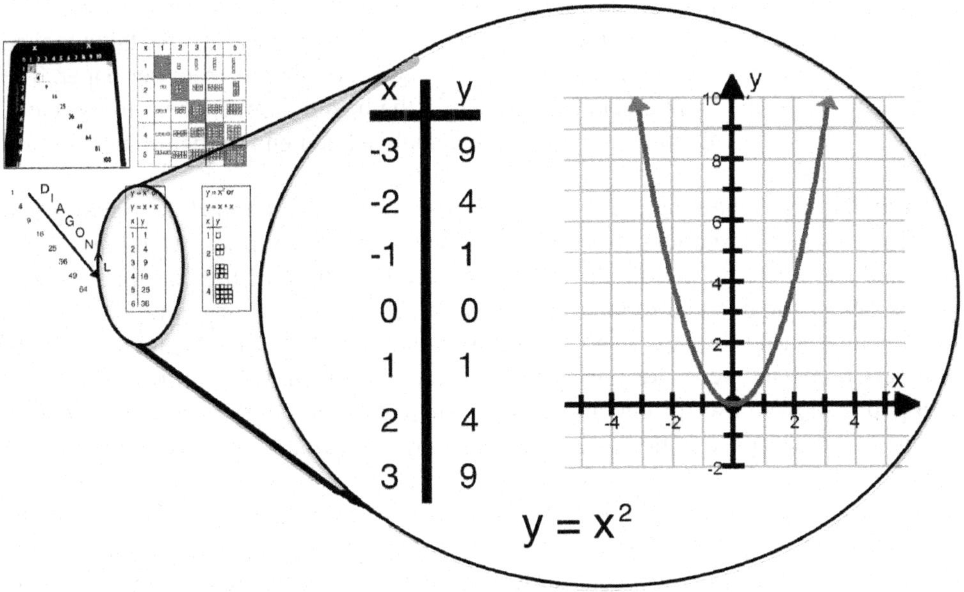

Figure 12.2. Diagonal multiplication table pattern from chapter 7 revisited for algebra.

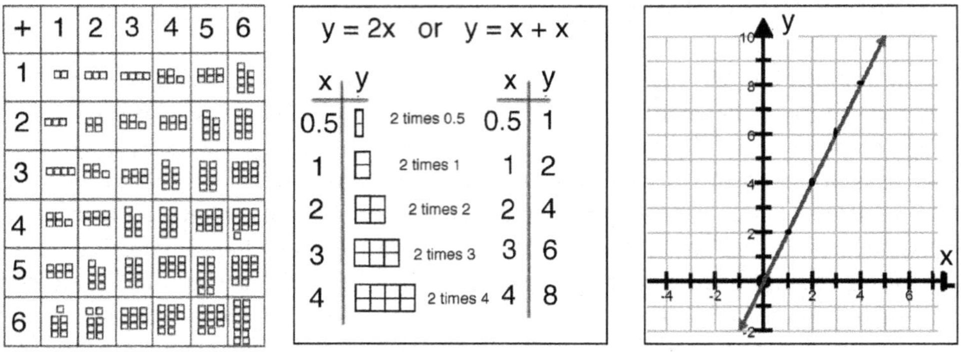

Figure 12.3. Diagonal addition table pattern from chapter 7 revisited for algebra.

After the algebra generalization into an equation happens, the numbers that could replace the "x" notation were again expanded to include 0.5 and could even reach into the negative numbers when the graph continues. The "x numbers" are always selected first because they were *here first* on that elementary number line. That elementary school number line was copied, rotated, and placed so that *both number lines crossed at zero*. This new vertical line earned another letter name "y" because "x" was already taken. Flexibility, as always, makes interpreting symbols in equations much easier to understand.

JILL'S GRAPHIC SOLUTIONS

Jill, a high school junior, used a subtraction strategy instead of square shapes to figure out one of the left-to-right diagonal patterns on the multiplication table in figure 7.3. The diagonal was just below the square pattern diagonal and the numbers were 2, 6, 12, 20, 30, 42, 56, 72, and 90. The order of these numbers in the algebraic table looked like the one in figure 12.4, along with her notes and subtractions from her paper. The replacements for the "x numbers" are shown in the boxed numbers next to the table. These "x" replacements in her equation are calculated and end up being the "y numbers" in the table.

After the last subtraction row, she had all twos so she stopped subtracting and started testing combinations. She knew that she would eventually graph these numbers and she knew it would help to know where the origin of the coordinate graph would be, so she decided to continue the subtraction pattern for numbers less than two by subtracting the next even number less than four. That led to another pair of numbers (zero and zero) for her table. The transition from the multiplication table to algebra shows how Jill decomposed the shapes in the table to show a repeating pattern, then she used numbers with that pattern and, finally expressed the pattern with "x" and "x^2."

Another pattern on a multiplication table that is *not in a straight diagonal* is the triangular number pattern. Flexibility is very helpful when searching for patterns because these triangular numbers appear in a step-diagonal instead of a straight diagonal. This triangular number pattern is 1, 3, 6, 10, 15, and so on, and the numbers appear

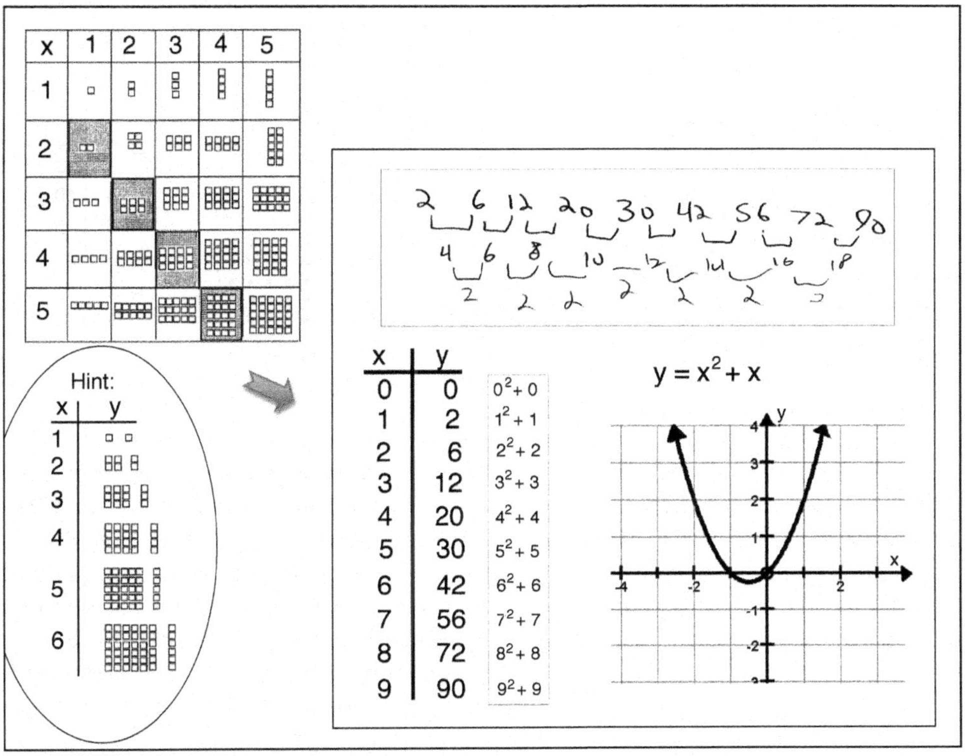

Figure 12.4. Jill's pattern solution and graph.

Try This:

Use the numbers in the pattern in Jill's algebra table from the y column. Try to arrange some shapes that could provide a clue about the equation. Here are some shapes to get you started.

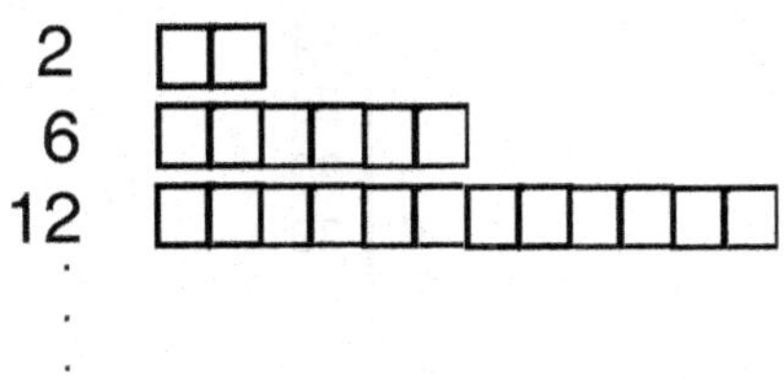

Arrange and rearrange the squares until you have convinced yourself that you have an equation that makes sense to you. Hint: Try making squares to see what is left.

Figure 12.5. Try This.

Multiplication Table

x	1	2	3	4	5	6	7	8	9	10
1	1	2	3	4	5	6	7	8	9	10
2	2	4	6	8	10	12	14	16	18	20
3	3	6	9	12	15	18	21	24	27	30
4	4	8	12	16	20	24	28	32	36	40
5	5	10	15	20	25	30	35	40	45	50
6	6	12	16	24	30	36	42	48	54	60
7	7	14	21	28	35	42	49	56	63	70
8	8	16	24	32	40	48	56	64	72	80
9	9	18	27	36	45	54	63	72	81	90
10	10	20	30	40	50	60	70	80	90	100

Triangle Numbers

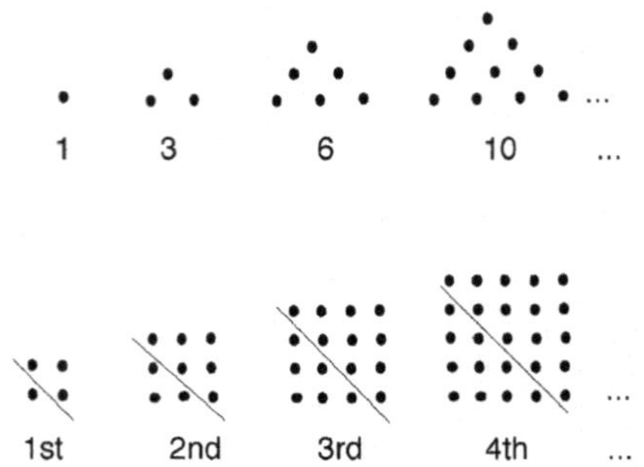

x	y
1	1
2	3
3	6
4	10
5	15
6	21

Answer: $\frac{n^2+n}{2}$ is the nth number

Figure 12.6. Triangle numbers and visual clue challenge.

in the shaded sections in the table on the left in figure 12.6. Arrangements of triangle dots that generate the numbers are in the center. An algebra table that lists the triangular numbers is located on the right side of the illustration. Arrangements of dots in squares offer clues to generate the equation for the pattern. Try to describe the pattern!

Some patterns lend themselves easily to shape arrangements and others are more difficult to see with shapes, especially the ones that include negative integers. Square and triangle areas are not measured with negative measurements but the patterns still exist, just as negative numbers exist. The more complicated or less-shape-driven arrangements are better strategized through algebraic reasoning.

Prove It to Yourself:

The numbers in this pattern come from a right-to-left diagonal on the multiplication table. Use shapes or Jill's subtraction strategy to prove to yourself that an equation will fit this pattern.

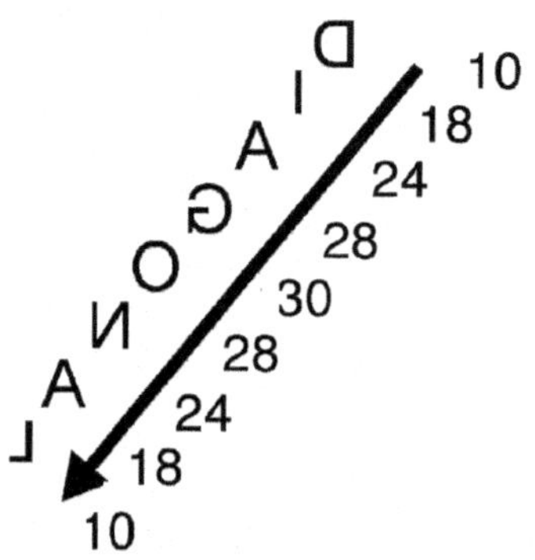

x	y
1	10
2	18
3	24
4	28
5	30
6	30
7	28
8	24
9	18
10	10

Hint: Make square shapes.
P. S. Don't "square" the "-" sign.

Answer: $-x^2 + 11x = y$

Figure 12.7. Prove It to Yourself.

FOIL'ED AGAIN

The process for multiplying algebraic binomials would look the same as that third- or fourth-grade multiplication procedure that you worked so hard to learn. Do you remember the acronym for multiplying algebra factors *F*irst *O*uter *I*nner *L*ast or *FOIL*. in your Algebra I class? *FOIL* is an algebraic mnemonic version for those multiplication problems from the third grade. Instead of 11 × 13 in figure 12.1, this example in figure 12.8 shows 45 × 53. Your child has already put in the time to learn that multiplication procedure for two-digit multiplication, so why not put it to good use again in an algebra lesson?

The *F*irst translates to the first terms in the two binomial factors (a binomial has two terms as seen in parentheses in figure 12.8). The *O*uter translates to the outer terms (the outside ends of the binomials or *one of the tens* and *one of the ones* in the traditional multiplication). Next, the *I*nner translates to a combination of the two inside terms of the two binomials or the tens numbers and ones numbers. Then, the *L*ast refers to both digits in the ones places. All of the diagrams that are in figure 12.8 represent the same idea. Find those *FOIL*ed binomials and make your way through the diagrams. The connecting clue: count the hundreds, tens, and ones pieces.

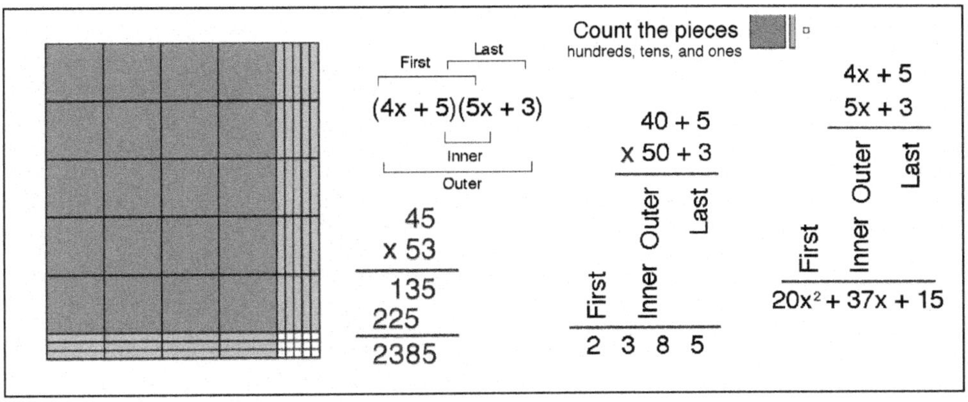

Figure 12.8. Five ways to see how *FOIL* works.

If counting is the clue, why have all of these different pieces? Different approaches are appropriate for different learners—visual-spatial and auditory-sequential. Visual learners appreciate knowing that the placements of those digits actually have a reason for their location in the problems. Auditory-sequential learners appreciate the sequence of *FOIL*. Since multiplication is commutative, the order of the multiplications (*LIOF* would work just as well) does not matter as long as the terms are *added* appropriately. It is really all about the blocks and place value and their related organization. Welcome back, Theon!

KEEP IN MIND

It is okay if you do not see the connections between arithmetic and algebra right away. Mathematics is *not* accomplished at Mach 3 speed. Real mathematics is about thinking, and thinking can, and should, take time. The thinking takes even more time if you or your child never had the time to think earlier. Visual learners must translate words to images and then back to words and that certainly takes more time! Finding connections, or just thinking about connections, can change the dreary "have to remember" part of math to being like other subjects "when you can think about it." The craft of mathematics is in the patterning and the strategizing.

NOTES

1. Saul, "Algebra, the Mathematics and the Pedagogy," 128.
2. Davis, *The Mathematical Experience*, 126.

Conclusion

What Parents Can Do

> Do not tell [your child] any rule. Let them experiment, see what they observe, and, when they are thoroughly familiar with the results, let them try to account for what happens.[1]
>
> —W. W. Sawyer

Tony's dad heard the noises his seven-year-old son was making and knew he was struggling. His son was trying to complete his math homework, a full page of mixed two-digit addition and subtraction problems with regrouping. Tony's dad had given Tony a five-minute "speed test" to try to motivate Tony to get through the page without distraction or procrastination. Tony understood regrouping, yet frequently found other things to do after completing about two or three problems. Tony was facing a *huge* mountain of pressure, but not about addition, subtraction, the switch between the two operations, or regrouping. The problem was "Mount Homework" practice pressure.

The sounds were disturbing, so after calming his son, he wrote a seven-digit subtraction problem complete with a "zero" or two peppered through the digits in the number. He gave the problem to Tony "as a challenge." Tony's dad wanted Tony to experiment with larger numbers. Tony accepted the challenge and began to subtract. After a minute or so, he proudly presented the *correctly* finished problem. Tony needed "proof" that "*he really did understand.*" He also needed to see *a rationale for* completing this mountainous page full of problems. Completing this million-long amount of digits in *one problem*, was "proof" to Tony *why* he was climbing Mount Homework. All those digits were coming in later problems.

Tony, a visual-spatial learner who was "King of Legos" at his house, had a difficult time staying focused on repetitive work. The mountain of problems became very tedious, not because he didn't know how to complete the work, but because it was not engaging and certainly not mentally challenging. This happens in many homes, sometimes because the child doesn't know how, sometimes because of other issues, and sometimes just because the task is too great to get through despite a child knowing how to complete the work. Full pages of practice problems are especially challenging, and ineffective, for visual learners.

This story also explains why listening beyond hearing and seeing what your child sees is so necessary. If your child is like Jackie, Trevor, or any of the other children in this book, then they have keen imaginations just waiting to be expressed, but not with Mount Homework practice. Sometimes motivation to focus is given with threat-like dictums, such as, "If you don't finish in five minutes, then I'll add three more problems." These kinds of reverse-motivation punishments can generate an even stronger dislike for math, *even if children like math now*. Like Tony, children can like math and have accomplished the skills, but their math enthusiasm can be dampened by tedium or extra work.

Avoid "falling in love" with a particular procedure just because it is the only one you know. If it *is* the only way you know, then you could fall prey to that "safety grip" on a procedure simply because "that's the way I learned it." Having a preference for a particular way of calculating after trying other ways is different than the "safety grip." Preference for a procedure is like preferring a particular sauce; you prefer it because the others did not appeal to your taste buds as much. Encourage your child to try some other "math sauces," create some of your own "math sauce," and then see if you can taste the differences.

PROBLEM SOLVING BEYOND WORD PROBLEMS

No one can know what problems will pop up in the future, so the best thing to do now is to *learn how to think* when solving any current problem. The average human brain has about one hundred billion neurons, each with about ten thousand connections for thinking thoughts. Imagine being able to use all of these connections to think about and answer questions that haven't even been asked yet. Word problems are stories, so enrich those stories with imagination. Trekkies or Ringers can start with science fiction; sports enthusiasts can use favorite sports information and players. Let your child engage their imaginations, then *they* can ask questions to analyze the information.

You do not need to know the "answers," but you do need to be willing to explore, ask, search, and question information so that your child can learn now how to think critically and continue this kind of thinking in their future years. George Pólya isn't the only mathematician and math educator who has suggestions for how to engage in the business of solving problems, so look for what others have suggested. Whether their proposals include steps or just guidelines, remember that all suggestions can help to generate ideas, especially your child's ideas.

As you pursue problem-solving strategies, make sure that the silence in the "wait time" for their responses is long enough to allow them to think. Remember, thinking requires time and patience; quick answers are just that, quick, and very often shallow. Quick answers can ward off pain, as Robbie thought. Problem solving requires making connections between pieces of information and associations among situations. Connecting and associating do not happen quickly. You be the judge. Is it heresy or just good sense to require that the mathematics make sense before requiring or encouraging any memorization?

Malika didn't know that decimals were connected to money. Fran didn't know that one-fourth as a fraction was like the quarter in a dollar. He didn't realize that he could figure out that 25 percent of anything was related to one-fourth or three-fourths, depending on how one thinks about the discounted sale price. "I was the kid who could get a decent, even good, grade on the tests without getting a deep understanding," wrote Joshua. Take the risk of asking questions that don't have obvious answers and explore "answers" with your child.

RIGHT TOOL FOR THE RIGHT JOB: A HARD LOOK AT TECHNOLOGY

Computers are to mathematics what telescopes and microscopes are to science.[2]

—Michael West

Ninety-six percent of respondents agreed to the statement: "Availability of calculators will permit treatment of more realistic applications of mathematics." This is not a quote from the twenty-first century. It was expressed in a 1974 math journal and has been repeated frequently in varied forms throughout the decades.[3] Technology doesn't inhibit mathematical thinking; technology enhances it. Calculations in arithmetic "service" mathematics and the other sciences and disciplines, a distinction that was first described in Plato's *Republic* with the age-old quadrivium (for those interested in history) outline of mathematics.[4]

The question needs to be "when to use technology," not "whether to use technology." Using technology influences how information is managed; it allows a student to change their approach from calculation to analysis and *from a problem statement to problem solving with ingenuity*. The engineers knew the calculated distance that the Apollo 13 spacecraft was from finishing the trip back to Earth, but it took ingenuity to literally make a "square peg fit with a round hole" when the oxygen connection was damaged.

Calculators can calculate very well when the correct keys are pressed in the correct sequence. What *calculators cannot do* is ask questions and *think about mathematics* and solve unusual problems. Calculators cannot press their own keys. Calculators cannot have conversations about thinking. Technology has not yet completely resolved how to do new thinking that is needed "to match squares with rounds." Calculators and computers make it possible for children to make errors and learn from those errors, whether the error is a consequence of *pressing the wrong keys or pressing the correct keys but in the wrong sequence.*

Different methods abound for calculating large numbers. The calculator makes all of those different methods obsolete, unless your child's finger hits the wrong key and they cannot tell if the answer on the screen is wrong. Calculators wrong? What a concept! Yes, calculators are very good at calculating, but your child must still do the thinking. If the calculator "operator" does not r*eally* understand what is going on, then the calculator is no longer helpful and, in fact, becomes a distinct hindrance. Calculators calculate so that a child can analyze the reasonableness of a calculator's results instead of spending time on repetitive calculations. Knowing more than one procedure puts your child in the driver's seat.

Sometimes math instruction software programs help students learn, and sometimes they don't. Some children need the hands-on physical work that a computer screen cannot provide. For these children, moving a curser isn't the same as moving a block.

The result, when it doesn't help, simply drives the Math Avoider deeper into confusion and frustration. Just ask Hilary and Jeannie.

Math software on a calculator or on a computer screen cannot "hear frustrated responses" or cope with ingenious solution options (unless the programmer has built it into the code). When only one or two alternative ways to "do the problem" are presented as "the only alternatives without taking into account what the child is thinking," then the results can be *really* emotionally devastating for the confidence-challenged learner.

Technology in all forms should be treated as math-supporting tools that enable real mathematics and the real communication of teaching, just as microscopes, telescopes, pencils, or those earlier quills dipped in white paint supported learning. There's a saying, "the best math happens on the back of an envelope or napkin." Neither calculators nor computers use napkins or envelopes. With such sophisticated calculators available, many a frustrated Math Avoider will wail about why anyone needs to learn any procedure at all. The craft of mathematics is in the design and the thinking notes on napkins that are written *before pressing keys on the calculator or computer.*

KEEP IN MIND

No one can predict what will be necessary to know in the future; what we do know is that we must know *how to learn.* Your child will be faced with the need to answer, and ask, questions about issues that will be happening in their future. They will need technology; they will also need to use critical thinking concerning their technology.

Don't allow your math fears to interfere with your child's math learning as confessed by Pulitzer Prize winning Eileen McNamara with, "I have doomed my eleven-year-old son's academic career. I didn't mean to do it."[5] Avoid letting your own perfectionistic tendencies interrupt your explorations with your child.

Now that you know that the mathematics itself didn't change in "new math," and that what you learned for long division still works, you can avoid frustration tears by helping your child "with the mathematics of action."[6] Work with the verbs *first* and then conform to the procedure structure later. Help your child maneuver through the newer math *procedures* by looking at them with "place value eyes." Making math make sense always takes priority. Dust off your own ingenuity and encourage your child's imagination and creativity. "Like the Grand Canyon, mathematics has its own kind of beauty and appeal to those who are willing to look."[7]

NOTES

1. Sawyer, *Vision in Elementary Mathematics*, 205.
2. West, *In the Mind's Eye*, 327.
3. NCTM Panel, "Where Do You Stand?" 488.
4. Turnbull, *The Great Mathematicians*, 11.
5. McNamara, *Boston Globe* (September 27, 1997).
6. Hogben, *Mathematics for the Millions*, 82.
7. Jacobs, *Mathematics, A Human Endeavor*, xiii.

Glossary

This glossary is generated from definitions provided by students. Please refer to a mathematical dictionary for formal mathematical definitions.

absolute value: distance between two number locations
additive inverse: zero, the "subtraction" to get to zero
algorithm: operation or procedure with numbers—add, subtract, multiply, divide, and all the other defined procedures
area: coverage inside of a shape
auditory-sequential: learns best by hearing step-by-step directions
circumference: the length around a circle space
circumscribed: when a polygon's corners touch the circumference of a circle
coefficient: the number in front of a variable or the number used as a base multiplier
commutative: the name given to the situation when the operation works both ways
complex number: a number that has a real number part and an imaginary number part
composition: building a term or shape with component pieces
congruent: the name assigned to shapes when their number measures are equal
coordinate points: a pair of numbers that shows where two number directions meet
cosine: the shift of a sine curve; the ratio of adjacent side to hypotenuse side in a right triangle
counting number: number that is used to count with. Also called a natural number
cylinder: a 3-D tube looking shape
decomposition: taking apart a term or a shape into component parts
diagonal: the one-dimensional line that connects two corners in a polygon
diameter: the one-dimensional line that connects the opposite ends on a circle circumference through the center
digits: 0, 1, 2, 3, 4, 5, 6, 7, 8, 9
directional reading: how your eyes track while reading something
distributive: when a number is multiplied with both numbers in an addition or subtraction parenthesis form
equal sign: a symbol that shows that two sides of an equation will balance

exponents: the small number in the upper right side of a term that tells how many times that term is multiplied by itself
factor: decomposing by only using division. No fractions
fractions: part-of-it as compared to all-of-it ratio
imaginary number: a number that does not belong to the Real number set. It has a $\sqrt{-1}$
infinity: a really big concept describing the idea of going on forever. It isn't a number but does have an aleph symbol assigned to it
inscribed: when a polygon is inside of a circle and the vertices of the polygon touch the circle's circumference
integer: a whole number that has either a + or a – indicator. No fractions
inverse operation: the opposite operation that takes numbers back to their "origin." For addition, the origin is 0; for multiplication, the "origin" is 1
irrational number: a number that cannot be written as a ratio and is usually written as an unending and non-repeating decimal
isosceles: the adjective that describes a triangle or trapezoid when two opposite sides are equal
Lattice method: an algorithm developed by John Napier
lightsaber: the long stretch of light in Star Wars and the name given to single long strip
logarithm: how to write a number that equals the exponent
natural number: a whole number but excludes zero
number sense: when numbers are used as abstraction ideas that only represent quantity
parabola: a symmetric curve that has an equation with the highest exponent of 2, like $y = x^2$
parallel: two lines or line segments that will never touch
parallelogram: a quadrilateral with opposite parallel sides
pentagonal number: a number that can make a pentagon shape
percent: a ratio comparison between a number and 100 and comes from place value
perimeter: the distance around the boundary of a shape
perpendicular: when a vertical line crosses a horizontal line
place value: one of the organizers for numbers that shows why numbers are written like they are
polygon: straight sided two dimensional shapes
polyhedra: straight sided three dimensional shapes
prism: many sided polyhedra
quadrilateral: four sided polygon
radical numbers: numbers that are written with a radical symbol like square root or cube root
radius: half of a diameter
rational number: a number that can be written as a ratio that compares two integers
reciprocal: how a number is written that can represent a multiplication inverse
regular: all sides have equal measures and all angles are equal to each other
rhombus: a quadrilateral with all adjacent sides congruent but not necessarily a square
scientific notation: a short way to write a very large or very small number using place value
similarity: same shape with proportional sides

sine: a continuous curve in trig; the curve goes through the origin (0, 0) and is the ratio between the opposite side and the hypotenuse in a right triangle
slope of a line: ratio between two points on a line comparing the distance up and over between these two points the ratio used as a coefficient to x in a linear equation; the ratio tells how high the line rises on y in relation to the run on the x axis between two points on the same line
square numbers: when a number of items can be arranged into a square
symmetry: when two sides of a shape exactly match
tangent: a noncontinuous curve in trig; the curve goes through the origin (0, 0) and is defined by the ratio of the sine over the cosine ratios of the sides of a right triangle
transcendental number: a number like pi or phi
trapezoid: a quadrilateral that has exactly two parallel sides
triangular number: a number that can make a triangle shape
trigonometry: a kind of math that studies ratios of sides in a right triangle that match to angles
Venn Diagram: a diagram with ovals that show how different groups are related to each other
vertex: a corner of a polygon or polyhedra
viewfinder: a piece of paper with cutout windows to put over tables in order to see patterns
volume: what fills up the contents of a 3-D shape
whole number: all of the counting numbers with zero included

References

Alsina, Claudi, and Roger B. Nelson. *Math Made Visual*. Washington, DC: The Mathematical Association of America, 2006.

Apleman, Maja, and Julie King. *Exploring Everyday Math*. Portsmouth, NH: Heinemann Publications, 1993.

Askey, Richard. "Knowing and Teaching Elementary Mathematics." *American Educator* (Fall 1999): 6–13. www.aft.org/sites/default/files/periodicals/amed1.pdf.

Baretta-Lorton, Mary. *Workjobs for Parents, Activity Centered Learning in the Home*. Menlo Park, CA: Addison-Wesley, 1975.

Boaler, Jo. *What's Math Got to Do with It? How Parents and Teachers Can Help Children Learn to Love Their Least Favorite Subject*. New York: Viking Press, 2009.

Boyer, Carl B. *A History of Mathematics*. Hoboken, NJ: John Wiley and Sons, 1968.

Brownell, William A. "The Progressive Nature of Learning Mathematics." *The Mathematics Teacher* 100 (January 2007): 26–34.

Cacha, Frances B. "Understanding Multiplication and Division of Multi-digit Numbers." *The Arithmetic Teacher* 19 (May 1972): 349–54.

Caliandro, Christine Koller. "Children's Inventions for Multi-digit Multiplication and Division." *Teaching Children Mathematics* 6 (February 2000): 420–26.

Common Core State Standards for Mathematics. Common Core State Standards Initiative, 2014. www.corestandards.org.

Connell, Maggie Martin. "3 Is Not Three—The Art of Digit Dancing." *Canada: SouthEast Education Network*, August 2009. www.seenmagazine.us/article/article-detail/articleid/5/math-moments-with-maggie.aspx.

Courant, Richard, and Herbert Robbins. *What is Mathematics?* Revised by Ian Stewart. New York: Oxford University Press, 1996.

Curcio, Fran, and Sidney L. Schwartz. "There Are No Algorithms for Teaching Algorithms." *Virginia Mathematics Teacher* (1998): 24–27.

Curico, Frances R., Francine Sicklick, and Susan B. Turkel. "Divide and Conquer, Unit Strips to the Rescue." *The Arithmetic Teacher* 35 (December 1987): 6–12.

Davis, Philip J., and Reuben Hersh. *The Mathematical Experience*. Boston: Houghton-Mifflin Company, 1981.

De Bono, Edward. *New Think*. New York: Avon Publishers, 1968.

Dienes, Zoltan Paul. *Mathematics as an Art Form*. (2002, 2004). www.zoltandienes.com/wp-content/uploads/2010/05/Mathematics-as-an-art-form.pdf.

Dienes, Zoltan P., and E. W. Harding. *Exploration of Space and Practical Measurement.* New York: Herder and Herder, 1966.

Dixon, John Philo. *The Spatial Child.* Springfield, IL: Charles C. Thomas, 1983.

Draper, Catheryne. *The Algebra Game: Linear Graphs, Quadratic Equations, Conic Sections, and Trig Functions*. Rowley, MA: Didax Education Inc., 2016.

Draper, Catheryne. "It's All about Connections." *Connects Magazine* (January–February 2006): 13–17.

Draper, Catheryne. *User-Friendly Math for Parents: Learning and Understanding the Language of Numbers Is Key*. Lanham, MD: Rowman & Littlefield, 2017.

Draper, Catheryne. *Winning the Math Homework Challenge: Insights for Parents to See Math Differently*. Lanham, MD: Rowman & Littlefield, 2017.

Eastaway, Rob, and Mike Askew. *Old Dogs, New Math: Homework Help for Puzzled Parents*. New York: The Experiment Publishing, LLC, 2010.

Enzensberger, Hans Magnus. *The Number Devil: A Mathematical Adventure*. New York: Metropolitan Books, 1997.

Escher, M. C. "M. C. Escher Quotes." AZ Quotes. www.azquotes.com/author/4555-M_C_Escher.

Eves, Howard. *An Introduction to the History of Mathematics*. New York: Holt, Rinehart, and Winston, 1969.

Fitzgerald, William M., and Jane U. Boyd. "A Number Line with Character." *Arithmetic Teacher* 41 (March 1994): 368–69.

Grandin, Temple. Thinking In Pictures. New York: Vintage Books, 2006. www.grandin.com/inc/visual.thinking.html.

Greenes, Carole E., and Rheta Rubenstein, ed. *Algebra and Algebraic Thinking in School Mathematics, Seventieth Yearbook*. Reston, VA: National Council of Teachers of Mathematics, 2008.

Halmos, Paul. "Mathematics as a Creative Art." The Royal Society of Edenburgh Year Book 1973 (Session 1971–1972). http://www-history.mcs.st-andrews.ak.ul/Extras/Creative_art.html

Hardy, G. H. *A Mathematician's Apology*. Cambridge, UK: Cambridge University Press, 1973.

Hersh, Reuben. *What Is Mathematics, Really?* New York: Oxford University Press, 1997.

Hogben, Lancelot. *Mathematics for the Millions*. New York: W. W. Norton and Company, 1951.

Hung-Hsi Wu, "What's Sophisticated about Elementary Mathematics: Plenty—That's Why Elementary Schools Need Math Teachers." *American Educator* (Fall 2009): 4–14.

Irving, Washington. *Tales of a Traveler*. New York: Century Company, 1909.

Jacobs, Harold R. *Mathematics, a Human Endeavor*. New York: W. H. Freeman and Company, 1982.

Kamii, Constance, "Arithmetic: Children's Thinking or Their Writing of Correct Answers." *The Arithmetic Teacher* 35 (1987): 2.

Kamii, Constance Kazuko. *Young Children Reinvent Arithmetic: Implications of Piaget's Theory*. New York: Teachers College Press, 1985.

Kamii, Constance, Barbara A. Lewis, and Sally Jones Livingston. "Children Inventing Their Own Procedures." *The Arithmetic Teacher* 41 (December 1993): 200–3.

Kasner, Edward, and James Newman. *Mathematics and the Imagination.* New York: Simon and Schuster, 1940.

Katterns, Bob, and Ken Carr. "Talking with Young Children about Multiplication." *The Arithmetic Teacher* 33 (April 1986): 18–21.

Kenschaft, Patricia Clark. *Math Power, How to Help Your Child Love Math, Even if You Don't.* New York: Addison Wesley Longman, 1997.

Liping Ma. *Knowing and Teaching Elementary Mathematics*. Hillsdale, NJ: Lawrence Erlbaum Associates, 1999.

May, Lola June, "Developing Multiplication Concepts." *The Arithmetic Teacher* 41 (April 1994): 446–48.

McKillip, William D. "Teacher Made Tests: Development and Use." *The Arithmetic Teacher* 27 (November 1979): 38–43.

McNeil, Ruth. "A Reflection on when I Loved Math and Why I Stopped." *Journal of Mathematical Behavior* 7 (1988): 45–50.

National Council of Teachers of Mathematics Panel. "Where Do You Stand? Computational Skill is Passe." *The Mathematics Teacher 67* (October 1974): 485–88.

National Council of Teachers of Mathematics. *Curriculum and Evaluation Standards for School Mathematics*. Reston, VA: National Council of Teachers of Mathematics, 1989.

National Council of Teachers of Mathematics. *Principles and Standards for School Mathematics*. Reston, VA: National Council of Teachers of Mathematics, 2000.

National Research Council. *Everybody Counts*. 1989. www.mathcurriculumcenter.org/PDFS/CCM/summaries/everybody_counts_summary.pdf.

Ornstein, Robert. *The Right Mind*. New York: Harcourt Brace and Company, 1997.

Paulos, John Allen. *Innumeracy*. New York: Hill and Wang, 1988.

Peterson, Penelope L., Elizabeth Fennema, and Thomas Carpenter. "Using Knowledge of How Students Think about Mathematics." *Chronicle Guidance* (January 1990): 23–27.

Philipp, Randy, and Alison Williams. *Helping Students Develop Confidence to Learn Mathematics*. A presentation at the California Mathematics Council, Southern Section, November 6, 2010.

Pólya, George. *How to Solve It*, second edition. Garden City, NJ: Doubleday Anchor Books, 1957.

Pólya, George. *Mathematical Discovery*, Volume I. New York: John Wiley and Sons, 1967.

Pólya, George. *Mathematical Discovery*, Volume II. New York: John Wiley and Sons, 1968.

Pólya, George. *Mathematics and Plausible Reasoning*, Volume I. Princeton, NJ: Princeton University Press, 1954.

Reinhart, Steven C. "Never Say Anything a Kid Can Say." *Mathematics Teaching in the Middle School* 5 (April 2000): 478–83.

Rubenstein, Rheta N. "Word Origins, Building Communication Connections." Reprinted from the *College Mathematics Journal*, in the *Virginia Mathematics Teacher* (1999): 37–41.

Saul, Mark. "Algebra: The Mathematics and the Pedagogy." *Algebra and Algebraic Thinking in School Mathematics, Seventieth Yearbook*. Reston, VA: National Council of Teachers of Mathematics, 2008.

Sawyer, W. W. *A Prelude to Mathematics*. Baltimore: Pelican Books, 1955.

Sawyer, W. W. *Mathematician's Delight*. Baltimore: Pelican Books, 1943.

Sawyer, W. W. *Vision in Elementary Mathematics*. Baltimore: Pelican Books, 1964.

Silverman, Linda Kreger. "Identifying Visual-Spatial and Auditory-Sequential Learners: A Validation Study." visualspatial.org/files/idvsls.pdf.

Silverman, Linda Kreger, and Buck Jones. *Upside Down Brilliance—Strategies for Teaching Visual-Spatial Learners*. Denver: DeLeon Publishing, 2002.

Skemp, Richard R. *Intelligence, Learning and Action*. New York: John Wiley and Sons, 1979.

Skemp, Richard R. *Psychology of Learning Mathematics*. Hillsdale, NJ: Lawrence Erlbaum Associates, Publishers, 1987.

Stein, Sheryn L. "Young's Vision." *The Mathematics Teacher* 86 (April 1993): 330–33.

Sutton, Suzanne. "Remedies for the Pain of Math Anxiety." *ENC Focus Review*. Washington, DC: US Department of Education Eisenhower National Clearinghouse (May 2004): 7–10.
Tammet, Daniel. *Thinking in Numbers*. Boston: Little Brown and Company, 2012.
Tobias, Sheila. *Overcoming Math Anxiety*. Boston: Houghton Mifflin Company, 1978.
Turnbull, H. W. *The Great Mathematicians*. New York: Barnes and Noble, 1993.
Twain, Mark . *The Adventures of Huckleberry Finn*. New York: Harper and Brothers, 1884.
Von Oech, Roger. *A Whack on the Side of the Head*. New York: Warner Books, 1983.
West, Thomas G. *In the Mind's Eye*. Amherst, MA: Prometheus Press, 2009.
West, Thomas G. *Thinking Like Einstein*. Amherst, MA: Prometheus Press, 2004.
Whitin, David J. "Becca's Investigation." *The Arithmetic Teacher* 41 (October 1993): 78–81.
Williams, G. Arnell. *How Math Works*. Lanham, MD: Roman & Littlefield Publishers, 2013.
Young, Jacob William Albert. *The Teaching of Mathematics in the Elementary and Secondary School*. New York: Longmans, Green, and Company, 1906.

MATERIALS

Addition tables
Algebra Game: Linear Graphs, Quadratic Equations, Conic Sections, and Trig Functions
Algebra tiles
Base ten blocks
Cardboard to make viewfinders
Creativity
Chips, disks, color counters
Geometry Sketchpad
Graphic organizers
Grid paper
Imagination
Linking cubes
Markers—colors
Multiplication table
One-inch blank or colored wooden cubes
Patty Paper (tracing paper)
Toothpicks, straws, chopsticks (word problem fence-building straight items)
Unifix cubes
Visualization
Weights (round plumber) for number balance
Yarn

About the Author

Catheryne Draper's tenure in math education has crossed the half-century mark. After graduating with a B.S. in mathematics and an M.Ed. degree in mathematics education and supervision from the University of Georgia, she worked as a high school teacher before serving as a district supervisor, state level advisor, professional development consultant, and math coach/teacher for both large urban and smaller public and private schools. She has taught math at all levels, from kindergarten through college. For almost thirty-five years, she has maintained an ongoing practice working with individual children and adults who "see" math differently.

She has worked as a math editor for textbook and supplementary materials in regular education and special education. She opened the Math Studio in the early 1980s to pursue her passion for providing visual and tactile instruction for all students. Her twenty-five-plus year tenure working with clients of the Massachusetts Rehabilitation Commission brought predictably successful results as she helped her clients succeed in their math learning goals.

As part of her work at the Math Studio, she developed and published *The Algebra Game Program*, an instructional and assessment card program for classrooms and individuals. In addition to teaching and working in educational publishing, Draper has written articles for, and contributed, to many journal publications about mathematics education pedagogy. She is the author of *Winning the Math Homework Challenge* and *User-Friendly Math for Parents* published by Rowman & Littlefield.

While in Georgia, Draper was instrumental in developing a K–8 instructional and assessment tool with a computer-driven assessment component and was responsible for bringing computer terminals into the district's high school curriculum. After leaving Georgia, Draper worked with national assessment companies and publishing companies prior to opening the doors of the Math Studio to further influence positive changes in mathematics education.

Lightning Source UK Ltd.
Milton Keynes UK
UKOW05n0112231117
313207UK00021B/1004/P

9 781475 834222